Seed Biology Updates

Edited by Jose C. Jimenez-Lopez

Published in London, United Kingdom

Seed Biology Updates
http://dx.doi.org/10.5772/intechopen.98120
Edited by Jose C. Jimenez-Lopez

Contributors
Rember Pinedo-Taco, Cecilia Figueroa-Serrudo, Leonel Alvarado-Huamán, Sercan Önder, Damla Önder, Sabri Erbaş, Murat Mutlucan, Muhammet Tonguç, Nikhat Farhana, Ripudaman M. Singh, Mohammed Gulzar Ahmed, Thouheed Ansari, Abdullah Khan, Ayesha Sultana, Treesa P. Varghese, Abdul Rahamanulla, Ashwini Somayaji, Rupa Das, Saikat Biswas, Mohammed Shu'aibu Abubakar, Attanda Muhammed Lawal, Tabi Kingsley Mbi, Tatah Eugene Lenzemo, Ntsomboh Godswill Ntsefong, Rodolfo López-Gómez, Essoh Aimé Césaire Elekou, Irene Perea-Arango, Luis María Suárez-Rodríguez, Jose C. Jimenez-Lopez, Andrea Roman-Mateo, Esther Rodriguez-de Haro, Jose M. Berral-Hens, Sonia Morales-Santana

First published in London, United Kingdom, 2022 by IntechOpen
IntechOpen is the global imprint of INTECHOPEN LIMITED, registered in England and Wales, registration number: 11086078, 5 Princes Gate Court, London, SW7 2QJ, United Kingdom

British Library Cataloguing-in-Publication Data
A catalogue record for this book is available from the British Library

Additional hard and PDF copies can be obtained from orders@intechopen.com

Seed Biology Updates
Edited by Jose C. Jimenez-Lopez
p. cm.
Print ISBN 978-1-80355-813-4
Online ISBN 978-1-80355-814-1
eBook (PDF) ISBN 978-1-80355-815-8

Meet the editor

Dr. Jose C. Jimenez-Lopez obtained a BS in Biochemistry in 1998, a BS in Biological Sciences in 2001, and an MS in Agricultural Sciences in 2004, all from the University of Granada, Spain. He obtained a Ph.D. in Plant Cell Biology from the Spanish National Research Council (CSIC) in 2008. During 2008–2011, Dr. Jimenez-Lopez was a postdoctoral research associate at Purdue University, USA. During 2012–2015, he was a Marie Curie Research Fellow (EU - FP7) at the University of Western Australia and CSIC. Currently, he is a tenured scientist at CSIC, working in the functionality, health benefits, and molecular allergy aspects of seed proteins from crop species of agro-industrial interest (mainly legumes). He is the author of more than 70 journal articles, 30 book chapters, 2 patents, and more than 130 international congresses. He is an active member of different scientific societies and an editor of multiple books.

Contents

Preface

Abiotic stresses such as salinity, drought, and temperature greatly affect seed physiology, plant growth, and plant development. Drought and heat stress represent some of the main abiotic constraints faced by plants. Any increase in the incidence and strength of these stressors, either individually or in combination, significantly decreases crop productivity, thus putting future global food security at risk.

One of the most critical stressors is temperature since it can cause huge economic and agricultural losses. The temperature has a great influence on seed physiological processes such as seed filling, embryo development, reserve compound synthesis and storage in the endosperm, and so on during the vegetative and reproductive stages.

Drought and heat stress significantly affect seed yields by reducing seed size, number, and quality. High temperatures can reduce the period of grain development, causing huge repercussions in grain filling and yield reduction in many crop plants. For example, the increase of 1 °C in average temperature can result in a 10% reduction in crop yield for species such as rice. Increased temperature also reduces crop quality. The extent of impairment during the reproductive phase of crop growth, largely affecting the seed-filling process, is a key cause of significant yield losses. Seed filling is influenced by some metabolic pathways, particularly the production and translocation of photoassimilates from leaves to seeds, such as importing precursors of reserves biosynthesis. The metabolic control of these processes is very sensitive to water deficiency and high temperatures due to enzymes and protein transports involved in metabolism and located in leaves and seeds. Moreover, the combination of stressors is tremendously unfavorable for seed yield and quality. This book highlights potential markers of target sites for regulating seed-filling events and how they are affected by abiotic stresses.

It also discusses primary dormancy, which is strongly influenced by environmental factors during seed development. Seed dormancy is controlled by several environmental factors such as temperature, light, and duration of seed storage. Understanding how these factors affect seed dormancy and germination during seed maturation is necessary for crop production. For example, an appropriate level of seed dormancy is needed to produce cereal crops. Humidity and low temperature throughout seed maturation are associated with a lack of seed dormancy in various cereal cultivars. The oilseed is affected by high temperatures during seed filling, which reduces seed dormancy. High temperature during early endosperm development leads to seed germination and seedling growth in rice.

This book presents comprehensive information on the genetic mechanisms and biochemical and physiological cues that govern seed filling, seed development features under stress environments, seed dormancy and germination, and agronomic management to help crops develop resilience to climate change.

Jose C. Jimenez-Lopez
Estacion Experimental del Zaidin,
Spanish National Research Council (CSIC),
Granada, Spain

Chapter 1

Seed Filling

Sercan Önder, Sabri Erbaş, Damla Önder, Muhammet Tonguç and Murat Mutlucan

Abstract

The synthesis of seed storage reserves occurs during seed filling, and many seeds contain large and characteristic levels of polymeric reserves. Storage reserves are found in the endosperm of cereal seeds and in the endosperm and/or cotyledons of dicot seeds depending of the plant crop species. Recently progress has been made in understanding the complex network of genetic regulation associated with seed filling. These advances in storage reserve quantity and nutrient quality contribute to a comprehensive understanding of reserve composition, synthesis, and regulation. Phytohormones such as abscisic acid (ABA), cytokinin, gibberellic acid, Indole-3-acetic acid (IAA), ethylene and their interactions play critical roles in seed filling and development. At different stages of seed development, the levels of different hormones such as ABA, IAA zeatin and zeatin riboside changes gradually from the beginning of the process to maturity. In addition, the quality and yield of seed storage reserves are significantly affected by the environmental conditions before and during the synthesis of the reserves. Given the fateful importance of seed storage reserves for food and feed and their use as sustainable industrial feedstock to replace dwindling fossil reserves, understanding the metabolic and developmental control of seed filling will be an important focus of plant research.

Keywords: early maturation, environmental factors, genetic regulation, plant hormones, storage reserves

1. Introduction

Seed development is divided into three stages: embryogenesis, which includes embryo development, early maturation, or seed filling, which includes the accumulation of storage reserves; and late maturation, which includes seed desiccation and the transition to dormancy. After seed filling and desiccation, seed longevity increases up to 30-fold and places the embryo in a dormant state. The seed filling period accounts for between 10 and 78% of the total seed development period, but its importance during seed filling is still overlooked. Seed filling is a crucial stage for all seed plants, involving the synthesis of carbohydrates, lipids, and proteins, as well as the mobilization and accumulation of various components in the developing seeds. Although the metabolic pathways responsible for the synthesis of storage molecules are well known, their regulation is not well understood. Although seed filling is under genetic control, these developmental processes are influenced by the environmental factors, such as heat and drought stress. Therefore, environmental factors have major impacts on the qualitative and quantitative characteristics of seed development and yield. Optimizing the rate and duration of seed filling could provide high and stable yields by reducing the potential negative effects of late maturation and maximizing the assimilation of metabolites formed by photosynthesis. Therefore, the rate and duration of seed filling are important determinants of seed quality and yield in many plants. In addition, understanding the complex processes during seed filling could help develop high-yielding cultivars under stress conditions.

In this chapter, we will attempt to summarize recent developments in the following areas: synthesis of storage reserves, genetic regulation, role of phytohormones, and effects of environmental factors to expand our understanding of these processes during seed filling. Given the emergence of new approaches to the study of seed filling and the tremendous growth of this topic in recent years, our discussion will inevitably be largely incomplete, and we apologize in advance to our colleagues.

2. Synthesis of storage reserves during seed filling

The main carbon source for biosynthesis and nutrient accumulation in seeds is sucrose, which is produced by the products of photosynthesis in plants. Sucrose is transported from the vegetative parts where photosynthesis occurs to the developing seeds. In cereals and legumes, nitrogen is transported to seeds mainly in the form of asparagine and glutamine; in some species, alanine may also serve as a nitrogen source [1]. Cerelas obtain nitrogen from organic or chemical fertilizers. Ureides, allantoin, and allantoic acids are the major forms of organic nitrogen transport in legumes [2, 3], and about 10–15% of organic nitrogen is transported as ureides in soybean and cowpea. Nitrogen-fixing symbiotic *Rhizobium* in nodules produce ammonium used for purine and uric acid synthesis, and uric acid is transported to neighboring uninfected cells to synthesize allantoin in peroxisomes [4].

During seed filling, carbohydrates are constantly transported from vegetative parts, so the conversion and accumulation of photosynthetic products varies greatly among plant species. In wheat and barley, the net photosynthetic activity in the flag leaf and spike is quite high, while the sugar produced in the leaves below the tassel makes a lower photosynthetic contribution to the grain in maize. However, sugars produced in the leaves enveloping the cob are efficiently transported to the developing seeds. In legumes, sucrose is deposited in the form of starch in the leaves and pods, and nitrogen is stored in the leaves and remobilized to the developing seeds.

Vascular tissues transport nutrients and water and terminate in the placento-chalazal region and seed coat of monocotyledons and dicotyledons, respectively. Nutrients are transported and released into the apoplast and taken up by the developing endosperm and embryo during seed filling. Vascular tissues across the developing grain facilitate the transport of nutrients to the endosperm in winter cereals. Specialized transfer cells facilitate the transport of nutrients from the pedicel to the endosperm in warm-season cereals. In legumes, reserves are accumulated in the cotyledons and nutrients are transported via vascular tissue to the funiculus, from where they enter the apoplast space and are then redistributed in the developing seed.

Seeds must have long-term energy stores in the form of starch, lipids, or hemicellulose to ensure successful germination and seedling development. These carbon sources are stored in the cotyledons or endosperms of most plant species. The conversion of assimilated carbon, usually in the form of sucrose, into various storage compounds in different tissues is regulated by complex interactions of gene expression and metabolic activity during seed development [5–7]. Starch is present in most plant tissues as a carbon storage compound and can account for up to 70% of the dry weight of seeds in many cereal grains [8].

The structure and composition of starch, which is inert and insoluble in water, makes it an ideal storage material that allows large amounts of sugars to be stored in cells without negatively affecting the dissolution potential in seeds. Starch accumulation begins in endosperm cells shortly after fertilization and rapid cell division [9, 10]. The number of endosperm cells can be used as an indicator of yield. The cell division phase is completed within 2–6 days after pollination (DAP), but cell volume continues to increase until maturity (~35–40 DAP) [11]. Accumulation of storage reserves usually occurs between 10 and 35 DAP [1].

2.1 Starch synthesis

There are at least four groups of enzymes involved in starch synthesis in plants. They are ADP-glucose pyrophosphorylases (AGPase), starch synthases (SS), starch branching enzymes (SBE) and starch debranching enzymes (DBE) (**Figure 1**). Plants usually have several isozymes of each group, 14 forms of these enzymes (2 AGPase, 5 SS, 3 SBE ve 4 DBE) are involved in starch synthesis and 13 of them show varying degrees of homology in all plans [13]. Sucrose is used as a substrate for starch formation to produce straight-chain amylose and branched amylopectin in seeds (**Figure 1A**). In cell cytosols, sucrose is converted to fructose (Fru) and uridine diphospho glucose (UDPG1c) by sucrose-UDP glucosyltransferase.

Fructose is phosphorylated by hexose phosphate isomerase to Glc-6-P and to Fru-6-P, which is converted to G1c-1-P by phosphoglucomutase. UDPG1c is also converted to Glc-1-P by UDPG1c pyrophosphorylase (UGPase). The first step of starch synthesis is the formation of ADP-glucose by AGPase [14]. The reaction catalyzed by AGPase is the first stable step in the biosynthesis of both temporarily stored starch in chloroplasts and chromoplasts and starch stored in amyloplasts. This enzyme is located in the plastids of photosynthetic tissue and has different forms in seeds, therefore its cellular location may vary in different plants. While most of the AGPase is found in the plastids of potato and pea, it is mainly located in the cytoplasm in maize, barley, and rice [15]. The enzyme carries out the following reaction: Plastid alkaline conversion of inorganic pyrophosphate (PPi) to inorganic phosphate (Pi) maintains the balance in favor of ADP-glucose synthesis through the action of inorganic pyrophosphatase [16], which can be transported through the plastid envelope [17]. The conversion of ADP-glucose occurs in the amyloplasts of storage cells of dicotyledons (**Figure 1**), while it occurs in the cytosols and plastids of endosperm cells of cereals (**Figure 1**). APGase is a heterotetrameric protein consisting of two large (APG-L) and two small (APG-S) subunits encoded by two different genes [18]. Plastidial AGPase is found in all starch-synthesizing tissues, but there are at least two different AGPases, corresponding to plastidial and cytosolic isoforms of AGPase present in the developing endosperm of maize [19], barley [20], rice [21], and wheat [22]. Starch synthesis by cytosolic AGPase depends on PPi-consuming reactions catalyzed by fructose-6-phosphate, 1-phosphotransferase, and UDP-glucose pyrophosphorylase for starch biosynthesis [23, 24]. The cytosolic AGPase isoform is responsible for 65–95% of the total AGPase activity in the developing cereal endosperms. Consequently, most starch biosynthesis occurs through the import of ADP-glucose in exchange for ADP, which is a byproduct of starch synthase in plastids [25, 26]. Starch biosynthesis in non-graminaceans depends on plastidial AGPase and the import of ATP and hexose phosphates from the cytosol (**Figure 1**). The forms and activity of APGases can vary in different parts of the

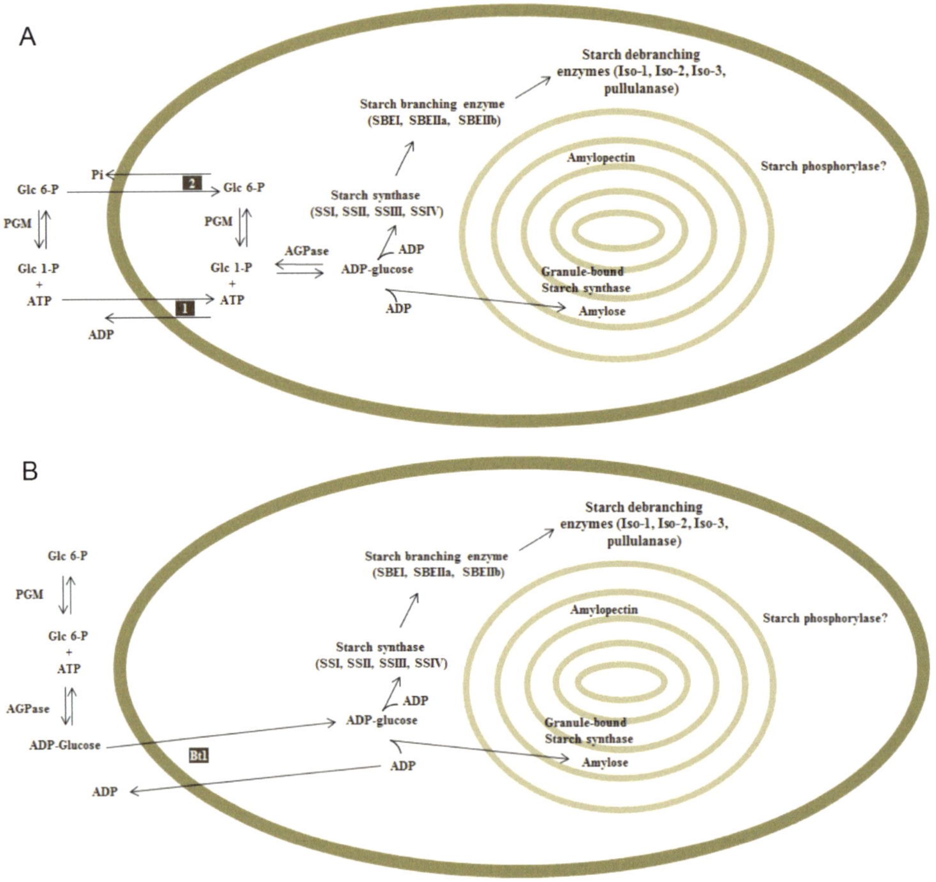

Figure 1.
Biosynthesis of starch in monocotyledons (A) and dicotyledons (B). (A) Monocotyledons have the cytosolic form of AGPase. ADP-glucose is taken up from the cytosol via the ADP-glucose/ADP transporter pathway (Bt1). (B) Hexose-phosphates and ATP are transferred from the cytosol into the plastid via the Glc 6-P/Pi antiporter (1) and the ATP/ADP transporter (2), which are located in the plastid inner envelope membrane. Cytosolic and plastidial isoforms of phosphoglucomutase (PGM) convert Glc 1-P and Glc 6-P into each other. Pi produces the pyrophosphate produced by AGPase. ADP is produced as a byproduct of starch synthase activity (SS). The starch synthases use ADP-glucose to form amylopectin with starch branching and debranching enzymes [12].

same plant because there are several genes encoding large and small subunits of APGases [27–29]. The APG-L subunits have specific expression patterns in different tissues, such as leaves, roots and endosperm of cereals [28, 30–32], or their expression levels are regulated under specific conditions, such as sugar content in potato [33, 34].

Starch synthases catalyze ADP-glucose to glucans and these are eventually used to synthesize water-insoluble amylose and amylopectin. Granule-bound starch synthases (GBSSI and GBSSII) produce and extend the amylose chain [35, 36]. Plants carrying a mutant form of GBSS corresponding to the waxy gene in cereals, do not produce amylose. They may also be involved in the elongation of glucans in various plants such as rice [37]. Another group of starch synthase enzymes (SSI-SSIV) is encoded by several genes and plays different roles in the formation of amylopectin in different tissues and developmental stages [38]. SSI produces short glucan (Glc) chains (<10). The first synthesized amylopectin is water-soluble and is elongated by SSII and SSIII to form water-insoluble amylopectin. Besides starch, there are other types of carbohydrates in the seeds of cereals and legumes, such as glucans and arabinoxylans. However, their amounts in the seeds does not exceed 10%, which is why they are not considered storage carbohydrates.

2.2 Lipid synthesis

Sucrose in seeds is used as a carbon source for the synthesis of triacylglycerols (TAGs). Lipid synthesis occurs in three steps: glycerol skeleton formation, fatty acid synthesis, and esterification of fatty acids and glycerol to complete lipid synthesis.

2.2.1 Glycerol synthesis

Glycerol synthesis reactions can occur in more than one way in plants, with the most common pathway being the fructose diphosphate (FDP) pathway. It is also known as the EMP pathway, after Emden-Meyerhof-Parnas, who discovered the pathway. Glucose is converted to fructose and enriched in energy by phosphorylation, then fructose-1,6-diphosphate is cleaved by aldolase to produce dihydroxyacetone phosphate (DHAP) and glyceraldehyde-3-phosphate (GAP), each with 3 carbons (**Figure 2**).

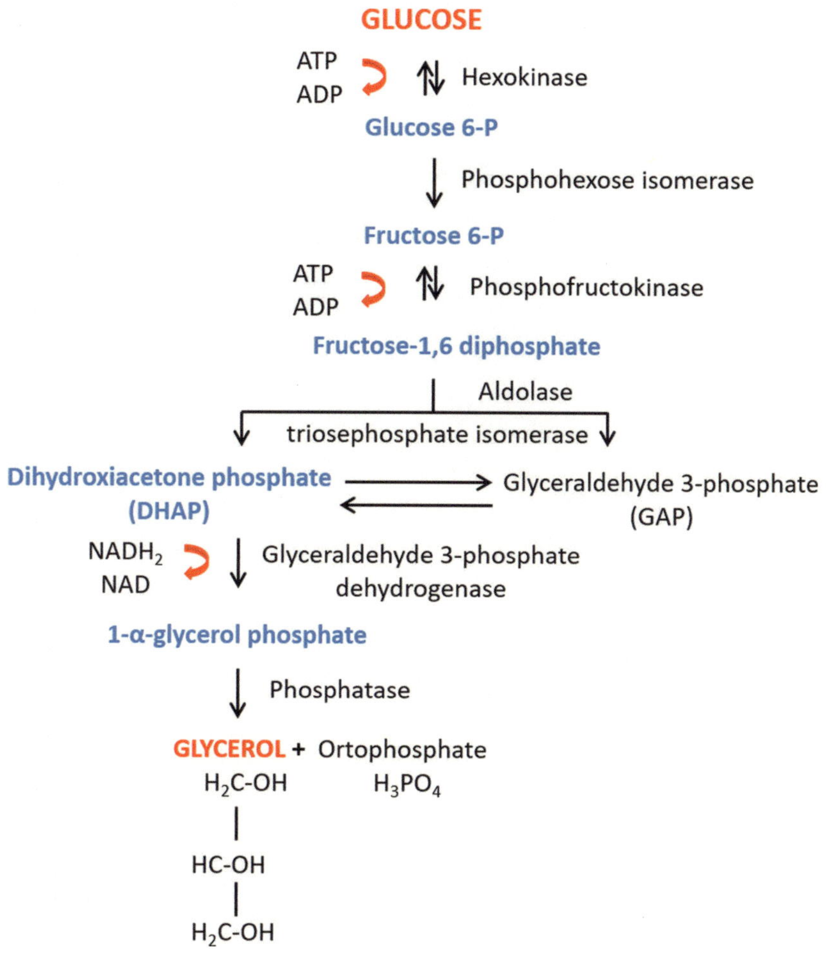

Figure 2.
Production of gylecerol from glucose [39].

The two triose-phosphates are in equilibrium with each other and the equilibrium is maintained by triose-phosphate-isomerase. DHAP is converted to glyceraldehyde phosphate by 3-phospho-glyceraldehyde dehydrogenase, and glyceraldehyde phosphate is further converted by phosphotosis to yield glycerol and orto-phosphate [39].

2.2.2 Fatty acid synthesis

Acetyl-CoA is used as a starting material for fatty acid synthesis. Glucose produced by photosynthesis is first converted to pyruvic acid and then to acetyl-CoA. Saturated fatty acids such as palmitate, stearate and oleate are produced from acetyl-CoA by acetyl-CoA carboxylase and fatty acid synthase (FAS) in the chloroplast of plants. The elongation of fatty acids to longer-chain fatty acids is catalyzed by elongases, and fatty acid desaturases (FAD) produce unsaturated fatty acids by inserting double bonds (desaturation) between carbon atoms (**Figure 3**) [40].

2.2.3 Lipid (triacylglycerol) synthesis

Lipids (triglycerides) are formed by combining a glycerol with three fatty acids via ester linkages. Glycerol and free fatty acids do not combine to form lipids, and glycerol-3-phosphate (G-3-P) and fatty acids bind to coenzyme A (fatty acyl-CoA) or acyl-bearing protein (ACP). The precursor of all fatty acids in seeds is acetyl-CoA, which is

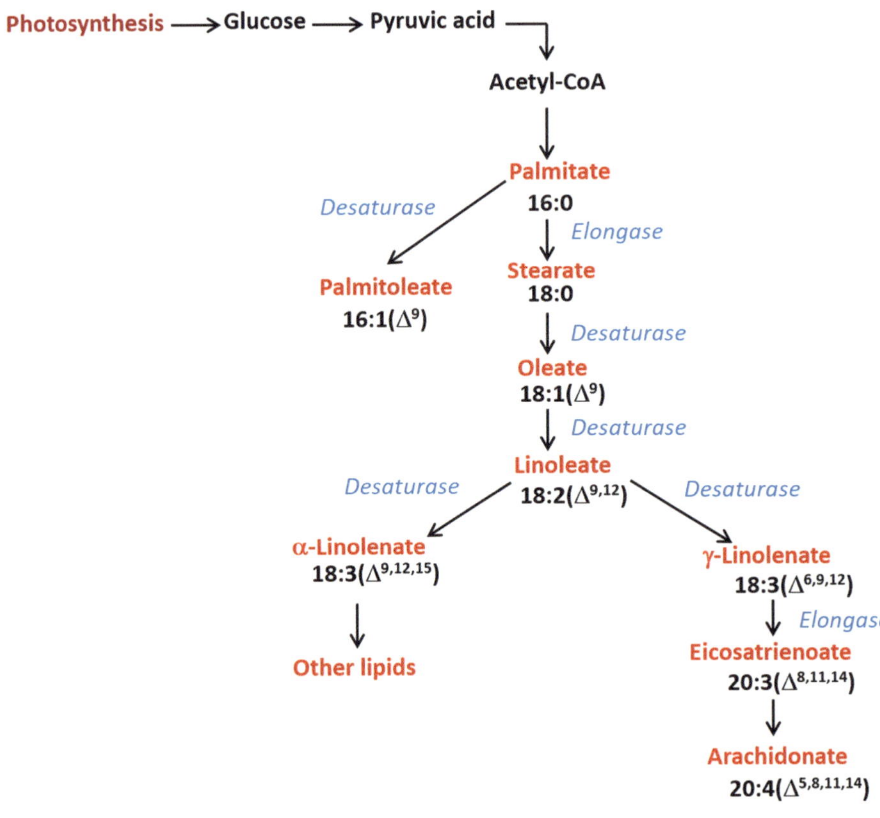

Figure 3.
Production of fatty acids from glucose in plants [40].

derived from sucrose in plants. After sucrose is transported into the developing seed, it is converted to hexose phosphate (Glc-6-P) and triose phosphate (Triose-P) in the plastids. The dihydroxyacetone phosphate is then reduced to yield glycerol-3-phosphate (G-3-P) in the cytosol. G-3-P is then esterified with three fatty acids in the endoplasmic reticulum to form triglycerides. Accumulation of G-3-P is an important limiting factor for the formation of new triglycerides. Glc-6-P is usually transferred to the plastids, but in some species it may also be converted to other intermediates in the cytosol or mitochondria. The first product for fatty acid synthesis is acetyl-CoA, which provides the 2C-acyl groups for the fatty acid chain. The first coupling step in fatty acid biosynthesis is the carboxylation of acetyl-CoA carboxylase (ACCase) to form malonyl-CoA. It is then converted to malonyl-ACP by addition of ACP with malonyl transacylase. Several enzymes are involved in a FAS complex that adds 2C to the extended chain and increases the length of the fatty acid at each cycle. Fatty acids are released from the ACP complex by acyl-ACP trioesterase (FAT) (**Figure 4**) [1, 40–44].

2.3 Protein synthesis

The synthesis and accumulation of storage proteins in seeds can vary due to genetic and environmental factors during seed filling. Synthesis of storage proteins occurs in specific tissues and depending on the stage of cell expansion during seed filling. During seed filling, endosperm, cotyledons, and embryo accumulate various proteins, including storage proteins, acid hydrolases, plant defense proteins and other reserve materials or metabolites. Vacuoles are the major storage organelles in seeds. Seed storage proteins undergo various modifications, including cleavage of signal peptides, glycosylation, folding, disulfide bond formation, and other proteolytic processes. These post-translational modifications are essential for seed protein functions (e.g., hydrolytic enzyme activity) and often have profound effects on seed protein accumulation in parenchymal storage cells (e.g., storage proteins). Proteins are transported simultaneously or shortly thereafter to the plastids, mitochondria, nucleus, and peroxisome/glyoxisome within plant cells by direct recognition through specific targeting signals. In contrast, transport to the vacuole and cell surface occurs through the endomembrane system with the endoplasmic reticulum, golgi complex, and transport vesicles. Proteins transported through the endomembrane system are synthesized on polyribosomes associated with the endoplasmic reticulum and first migrate into the lumen and enter the secretory system. In addition to seed storage proteins, other seed proteins are also transported via the secretory pathway during seed development, including those destined for the tonoplast (vacuolar membrane), plasma membrane, and cell wall matrix [45].

2.3.1 Cleavage of signal peptides

In most cases, the signal peptide is cleaved from the nascent polypeptide chain upon leaving the endoplasmic reticulum by a protease (signal peptidase) located on the inner surface of the membrane. The cleavage usually occurs in the C-terminal region, which allows further processes such as folding and assembly of the protein [46].

2.3.2 Protein folding and assembly

In order to have a three-dimensional structure and function, all proteins must undergo protein folding. With few exceptions, protein folding and assembly occur in the lumen of the endoplasmic reticulum and contribute to protein stability and efficient

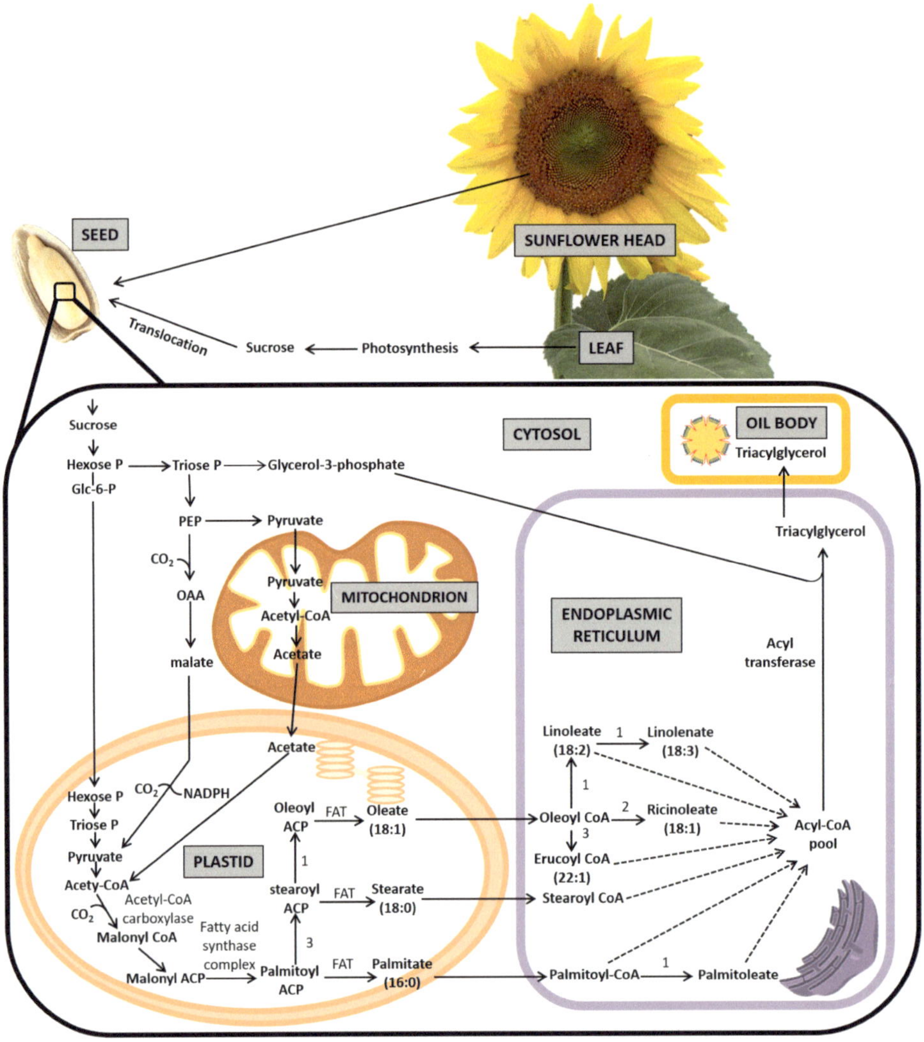

Figure 4.
Synthesis of fatty acids and oils in intracellular organelles during the seed filling period after fertilization. 1: FAD enzymes, 2: hydroxylase enzymes, 3: elongase complex enzymes, FAD: fatty acid desaturases, Glc-6-P: glucose-6-phosphate, PEP: phosphoenol pyruvate, OAA: oxaloacetate, ACP: acyl carrier protein, CoA: coenzyme A, FAT: fatty acyl thioesterase [1].

transport [1, 45–49]. Many plant vacuolar proteins, including storage proteins are oligomers, most of which are oligomerized in the endoplasmic reticulum. Chaperone proteins in the endoplasmic reticulum promote proper folding of polypeptides, and protein disulfide isomerase (PDI) forms disulfide bonds within and/or between peptides using the -SH groups of cysteines. The function of chaperones depends on their ability to recognize a variety of nascent polypeptides that do not share unique similarities, while accurately discriminating between properly folded and unfolded structures [50].

2.3.3 Glycosylation

Glycosylation is a commonly used modification of vacuolar proteins in seeds. The asparagine residue is used for N-linked glycosylation, and vacuolar glycoproteins usually have both high mannose content and complex N-linked glycans [46, 48, 49]. This process occurs in the lumen of the endoplasmic reticulum because the polypeptide is still in translation and contains a lipid carrier molecule (dolichol phosphate). There are two main types of oligosaccharide side chains: simple or mannose-rich oligosaccharides, which consist of mannose and N-acetyl-glucosamine, and complex or modified oligosaccharides, which are usually rich in mannose but also contain other residues such as fucose, xylose, and galactose. These oligosaccharide side chains may be found on the same polypeptide [45]. Glycosylation is thought to increase the stability of proteins and assist them in folding and assembly [1].

2.3.4 Proteolytic cleavage

Seed proteins transported by the secretory pathway are often subject to post-translational proteolytic cleavage, including storage proteins, seed defense proteins, and various vacuolar enzymes (α-mannosidase and thiol proteases). For seed storage proteins, the process begins during the transition to the vacuole and is completed in the vacuole [51]. Polypeptides can be split into two or three smaller peptides, some peptide chains can be removed, and the N- or C-terminus of peptides can be truncated [51, 52]. These modifications, together with glycosylation, could lead to a heterogeneous pool of mature proteins derived from a single polypeptide. Among storage proteins, 2S albumins undergo the largest posttranslational modifications [1, 45].

2.3.5 Formation of disulfide bonds

Many seed proteins must have disulfide bonds to stabilize their tertiary and quaternary structures. Disulfide bond formation occurs through disulfide isomerase, which promotes disulfide formation, isomerization or reduction, in newly formed proteins [53]. The enzyme interacts with unfolded proteins in the endoplasmic reticulum and catalyzes thiol oxidation and disulfide exchange reactions.

3. Genetic regulation during seed filling

Filling occurs when embryonic development is complete in the seed. Seed filling is under genetic control and is tied to changes in storage reserves. Seed filling is regulated by a network of signals mediated by various hormones [54, 55]. Basic research and information on differentiation, growth, and signal transduction related to seed filling and development is derived from studies in *Arabidopsis thaliana*, and there are sufficient reports to suggest that the major mechanisms regulating filling and development are similar in all plant seeds [55–57]. Recent developments in molecular genetics and genomics have led to a better understanding of the processes that occur during seed filling and development and offer opportunities to control and modify seed quality as well [58]. In addition, understanding the genetic factors that influence seed development could help breeders manipulate filling rate and duration to obtain higher yielding varieties.

Plotting phenotypic values on the growth curve throughout the duration of seed filling and analyzing the results using quantitative genetic approaches is a good strategy for exploring a time-dependent trait [59] to understand seed filling. Gene expression regulating storage reserves during seed filling is interrelated [54, 55, 60] and many known genes are expressed in the endosperm of flowering plants (**Table 1**). In *Arabidopsis*, four major regulators (ABSCISIC ACID INSENSITIVE3 [ABI3], FUSCA3 [FUS3], LEAFY COTYLEDON1 [LEC1], and LEC2) control many aspects of seed development, such as the accumulation of storage molecules, cotyledon identity, and the transition to desiccation tolerance and dormancy [66]. The ABI3, FUS3, LEC1, and LEC2 network of regulators has the common phenotypic effect of reduced expression of seed storage proteins. In addition, the main role of the LEC2 regulatory network is to up-regulate *FUS3* and *ABI3*. In *Arabidopsis*, several genes have been identified, such as the *FERTILIZATION INDEPENDENT SEED* (*FIS*) genes *MEDEA* (*MEA*) [67, 68], *FERTILIZATION INDEPENDENT ENDOSPERM* (*FIE*) [69], *FIS2* and *MULTI-COPY OF IRA1* (*MSI1*) [70, 71], *MEA* homologs *CURLY LEAF* (*CLF*) or *SWINGER* (*SWN*) [72], and *MATERNALLY EXPRESSED PAB C-TERMINAL* (*MPC*) [73] and *FLOWERING WAGENINGEN* (*FWA*). These genes are all involved in seed filling and early seed development in plants [74, 75].

DNA methylation is also one of the first recognized epigenetic modifications that affect gene expression by determining chromatin structure and compartmentalization of DNA during seed filling [55, 76]. *METHYLTRANSFERASE1* (*MET1*) is the major methyltransferase gene in *Arabidopsis* [77] and DNA methylation by *MET1* is involved in epigenetic control of seed size [78]. Transcriptome dynamics during seed filling have been described in several crops. As observed in barley and wheat seeds, the transition from cellular differentiation to filling in rice seeds is associated with

Plant	Gene full name	Gene symbol	Proposed gene function/ description	References
Pisum sativum	*TRYPTOPHAN AMINOTRANSFERASE RELATED2*	*TAR2*	Expression of the auxin biosynthesis gene (TAR2) is promoted by trehalose 6-phosphate, and affects auxin concentration by mediating the activation of storage processes in seed filling	[61]
Glycine max	*WRINKLED1*	*WRI1*	Regulates fatty acid accumulation and hormone signaling	[62]
Zea mays	—	*ZmABI19*	Acts as a seed filling start regulator. Also, involved in the accumulation of starch and fatty acids, and in signal transduction of plant phytohormones	[63]
Oryza sativa	*SUPER STARCHY1/ ONAC025*	*SS1/ONAC025*	Seed-specific gene in rice that promotes seed filling and adversely affects vegetative growth	[64]
Triticum urartu	Storage protein repressor	*SPR*	TuSPR expressed in endosperm during seed filling, suppresses the synthesis of seed storage proteins	[65]

Table 1.
Some genes involved in seed filling regulation.

changes in gene expression patterns [79, 80]. Using microarray technology, more than 20.000 genes associated with seed filling have been identified in rice, many of them related to metabolic pathways of carbohydrates and fatty acids [81]. The results of cluster and correlation analysis of these genes revealed 269 genes associated with seed filling [81]. In alfalfa, cluster analysis identified 5165 genes involved in seed filling, and most of these genes were associated with metabolic pathways of proteins for seed storage [58, 82]. The major regulators of gene expression are miRNAs and their expression has been studied in some crops [83], including rice, wheat, and maize [83–85]. Members of the miR156 family are specifically expressed during seed filling in rice by targeting genes of the squamosal promoter-binding protein-like (*SPL*) family. One of these genes, *SPL16,* controls cell proliferation during seed filling, and increased expression of *SPL16* correlates positively with grain yield in rice [86]. miR397 enhances brassinosteroid signaling by down-regulating the *laccase* (*LAC*) gene, which increases grain yield in rice [87]. Also, the expression levels of miR156, miR164, miR166, miR167, and miR1861 suggest that they play regulatory roles in rice during seed development and filling [84].

The major storage reserves accumulated during seed filling are storage proteins, lipids (generally TAGs), and carbohydrates (generally starch). Regulatory networks controlling seed filling are repressed prior to germination to prevent the accumulation of reserves during the vegetative development. Therefore, studies of gene expression during seed filling in tissues at the vegetative developmental stage may provide insight into the regulatory mechanisms underlying seed filling. The transcription factor WRINKLED1 (WRI1) plays an important role in fatty acid accumulation during seed development in *A. thaliana* [88]. Genetic and molecular studies suggest that WRI1 is a target of *LEC2*, *ABI3, FUS3*, and *LEC2* to regulate oleosin expression and lipid accumulation [88]. In *Arabidopsis*, mutations in *LEC1, LEC2* and *FUS3* resulted in decreased accumulation of storage proteins and TAGs [89, 90]. Synthesis of fatty acids in lipid metabolism during seed filling occurs through stimulation of fatty acid synthase or acyl carrier protein genes [58].

Sugar molecules can act as signaling molecules that regulate genes expressed in photosynthesis and metabolism. High sugar content promotes starch biosynthesis, while it has a negative effect on photosynthesis. Low sugar content increases the expression of genes related to photosynthesis and promotes the transport of seed reserves, while it limits the metabolic processes of carbohydrates [91]. Sucrose content controls cell differentiation and filling processes in seeds by altering gene expression and enzyme activities [92]. In faba bean, pea, and barley seeds, sucrose initiates gene expression regulating seed storage reserves and triggers the transition from embryogenesis to seed filling [93, 94]. Sucrose induces gene expression of globulin and albumin proteins, and *LEC1*, *LEC2* and *FUS3* are important regulators of sucrose in *Arabidopsis* [95]. Sucrose is also imported and converted to starch in endosperm during seed filling [96]. In *Arabidopsis*, mutations in sucrose transporter gene (*AtSUC5*) delayed the conversion of sugars to lipids, and the *AtSUC5* gene is involved in seed filling [97]. In rice, *AGPS2/shrunken2* (starch synthesis gene) is upregulated during the period of increasing seed dry weight [98]. Two rice sucrose synthase (*SUS*) genes are expressed, one at the early stage of seed filling and the other during seed filling, but not in the endosperm [81]. The *GRMZM2G391936* and *GRMZM2G008263* genes are involved in starch and sucrose metabolism during seed filling in maize [59]. The gene *GRMZM2G008263* is the starch synthase gene responsible for the production of amylose and is found only in starch grains [59]. The *GRMZM2G391936* gene encodes the large subunit of ADP-glucose pyrophosphorylase (AGPase). Alteration of AGPase activity can increase the yield of starchy plants [99–101]. *Trehalose-6-phosphate synthase1* (*tps1*) mutants demonstrated the importance of sugar signaling molecules during seed filling by down-regulating genes for starch-sucrose degradation and up-regulating genes for lipid mobilization to produce glucose [102]. Therefore, sugars as signaling molecules are important regulators during seed filling.

The amino acid content and the composition of the seed storage proteins influence the nutritional value of the seeds. Storage proteins are synthesized during seed filling and deposited in endosperm tissues. The rate of amino acid synthesis controls the rate and yield of storage protein synthesis. The phosphoenol pyruvate carboxylase (PEPC) enzyme is a critical factor in the biosynthesis of storage proteins in soybean, pea, faba bean, and wheat. Therefore, PEPC can be used to increase the protein content of seeds. Overexpression of PEPC in bean seeds results in up to 20% higher protein content per gram dry weight due to increased sugar/starch and free amino acid content [103], which led to the identification of an important marker for the transition from seed filling to the drying stage. Up-regulation of genes involved in amino acid metabolism (such as the amino-transferase gene) during seed filling in alfalfa results in increased amino acid synthesis, which is required for the production of seed storage proteins [58]. In maize, the expression level of marker genes for amino acid synthesis during seed filling has been studied [104]. One of these genes, *ZmAS1,* was expressed in both cobs and kernels, while others, *ZmAS2* and *ZmAS3,* were expressed in kernels. In alfalfa seeds, most of the storage reserves accumulate between 14 and 36 DAP in the embryo at the seed filling stage [105]. The *stress-associated protein 1* (*MtSAP1*) gene of alfalfa directly regulates the accumulation of seed storage proteins [106]. Phaseolin is the major seed storage protein in bean and the phaseolin (*phas*) gene is not expressed during the vegetative phase of plant development [107].

Genes responsible for the accumulation of storage proteins and lipids during seed filling are controlled by cis-acting elements in promoters. Well-characterized cis-elements are the RY repeats (CATGCA), the ACGT box (CACGTG), and the AACA motifs, controlled by the B3, bZIP, and MYB domain transcription factors, respectively [55]. For example, silencing of the *phas* gene in vegetative tissues has been associated with the presence of TATA boxes in the *phas* promoter [108].

Abscisic acid (ABA) is a key hormone involved in the regulation of several processes of seed development, such as maturation and reserve accumulation [109]. In *Arabidopsis*, barley and bean seeds *CYP707A* genes regulate ABA degradation in the embryo and endosperm [110–112]. In addition, gibberellins (GA) and ABA are also involved in cell differentiation and grain filling processes [112, 113]. While the level of GA is suppressed during seed filling, the level of ABA increases. In *Arabidopsis*, the biosynthesis of GA is controlled by the expression of the *AtGA2ox6* gene, but its expression is controlled by ABA levels [114]. Auxin is also involved in the seed filling process and interacts with the transcription factors LEC2 and FUS3 [55, 115]. The transcription factor ABI3 is involved in auxin signaling [116]. Expression of *LEC2* activates auxin-related genes [117] and auxin activates the expression of *FUS3* [118].

Flavonoids such as proanthocyanidins and anthocyanins are accumulated in seeds during seed filling. In alfalfa, several genes such as *MtWD40-1* [119], *MtMYB5* and *MtMYB14* [120], *MtPAR* (a regulator of proanthocyanidin accumulation through its effect on *MtWD40-1*; [121]) have been identified to be involved in the proanthocyanidin biosynthetic pathway during seed filling. In addition, genes for flavonoid biosynthesis have been identified, including chalcone synthases (*Mtr.42237.1.S1_at*), chalcone isomerases (*Mtr.40331.1.S1_at*), and flavonol synthases (*Mtr.45897.1.S1_at*) in alfalfa [58]. These enzymes cause the accumulation of isoflavones in the embryo

[122], but may also be involved in the accumulation of proanthocyanidins in the seed coat and tannins in the bark tissues [123]. Other genes involved in the accumulation of proanthocyanidins have been identified, such as glycosyltransferase (*UGT72L1*) and the proanthocyanidin transporter *MATE1* [124], which is responsible for the synthesis and modification of proanthocyanidin precursors [125]. *MATE2* transports anthocyanin by diverting flavonoid precursors into the anthocyanin pathway [126]. In addition, glycosyltransferase (*UGT78G1*) is required for the modification and accumulation of anthocyanins [127]. All of these genes are involved in the control of anthocyanidin reductase, one of the major enzymes responsible for the production of proanthocyanidins.

4. Role of phytohormones during seed filling

Seed development is divided into two main phases: the cellular phase and maturation [128]. The cellular phase includes all the processes involved in the formation and development of the different parts of a seed. In this stage, storage reserves for the embryo are synthesized and seed filling takes place. Phytohormones regulate signaling between the embryo and the endosperm. Most studies on seed filling and development have used *Arabidopsis* and maize as model plants for dicotyledons and monocotyledons. Although monocotyledonous and dicotyledonous plants share common seed characteristics, seed filling and developmental processes differ significantly between the two groups. In developing seeds, precise coordination is required to organize cell distribution in tissues and organs, and to control seed filling. The cells in the seeds can control all these activities by producing and sensing signals. The synthesis and regulation of phytohormones in the process of seed filling is essential for seed development [129, 130]. Seed filling is a highly coordinated and complicated process involving hormonal control and constant exchange of signals between different parts of the embryo [128].

Many studies have shown that hormone levels change during seed development and filling. Phytohormones, such as ABA, GA, cytokinins, Indole-3-acetic acid (IAA), and ethylene regulate seed filling processes (**Figure 5**) [132, 133]. Phytohormone gradients are synthesized in distinct seed sections, and their ratio controls signals that activate or inhibit specific seed filling processes. Among the hormones, ABA plays a central role as it accumulates at high levels from fertilization to seed maturation. Therefore, ABA functions as a signaling molecule and is important for seed filling, seed growth, dormancy, and plant stress responses [134]. Seed filling rate was positively associated with the concentration of ABA, and higher concentration of ABA resulted in higher seed filling rate [135]. In maize, the concentrations of ABA were associated with seed filling rate and kernel weight [136, 137]. Seed filling in barley, wheat, rice, and sorghum is closely related to senescence and the senescence-related hormone ABA, which affects nutrient mobilization and grain filling time [138] and is involved in the expression of senescence-related genes in barley [139]. High ABA levels increase remobilization of previously stored carbon in grains and accelerate grain filling rate [140] and have significant effects on seed filling in upper and lower grains [141]. ABA also inhibits cell cycle while accelerating seed maturation by upregulating inhibitors of cyclin-dependent kinases, which are important regulators of the cell cycle [142, 143]. While ABA has a positive effect on stomatal activity, seed dormancy, and plant response to abiotic and biotic stresses, it has a negative effect on seed germination [144]. Other plant hormones, such as gibberellins, ethylene, cytokinins,

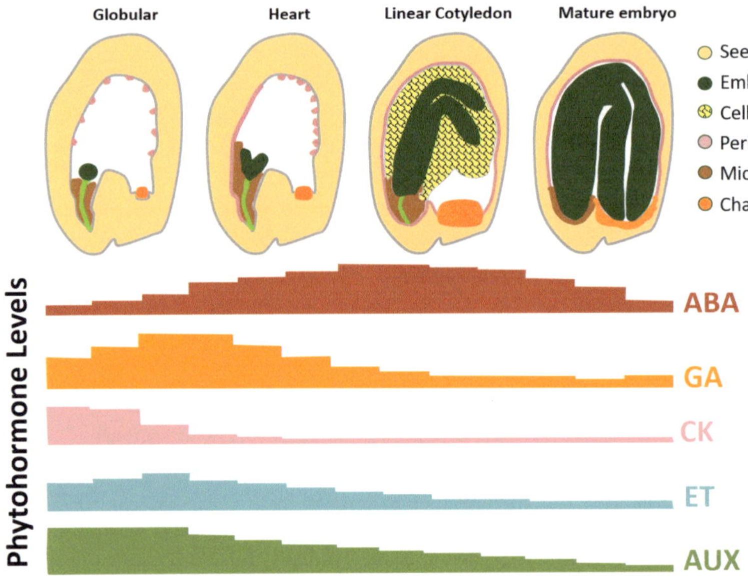

Figure 5.
Schematic representation of phytohormone accumulation during seed development. (A), represents the stages from late seed development to seed maturity of a dicot plant. (B) Shows the changes in specific phytohormone levels from top to bottom. Longer bars indicate higher levels. Three types of endosperm are formed during maturation: unicellular stratified endosperm, micropylar endosperm, and chalazal endosperm. High concentrations of abscisic acid, present at all stages of seed development, are thought to play a key role in seed filling. Gibberellins are synthesized in the differentiation stage of the embryo to promote cell growth and expansion, and in the late maturation stage to activate proteolytic enzymes. Accumulation of abcisic acid inhibits all processes induced by gibberellins. The accumulation pattern of cytokinins is opposite to that of abscisic acid. Cytokinins play an important role in cell division and their levels gradually decrease during the cell enlargement phase. Ethylene production increases during the early stage of seed development. Auxin controls grain filling by regulating invertase activity. An increase in auxin levels improves sink capacity and nutrient uptake [131].

brassinosteroids, and their antagonistic interactions with ABA could improve seed germination. ABA can stimulate sucrose storage in the seed coat and accelerate sucrose transport to the cotyledons during seed filling [145]. Gibberellins are also involved in cell differentiation and seed filling. Gibberellin concentration in seeds was not significantly related to seed filling rate or seed weight [146], but GA content had a negative effect on seed filling rate in rice seeds [140]. These studies showed that ABA and GA have antagonistic effects during seed filling [147]. The amount of bioactive GA decreased at the stages when ABA peaked with inactivation reactions to ensure normal seed filling and growth [148].

Cytokinins are involved in cell division, chloroplast formation, senescence, and stress tolerance in plants [149]. Cytokinins also play an important role in seed filling by inducing rapid cell division of endosperm cells [131]. In addition, zeatin (Z) and zeatin riboside (ZR) are biologically important cytokinins in higher plants [150]. Zeatin and ZR contents increase fertilization, kernel set, and endosperm growth in cereals [151]. High Z and ZR contents are necessary for seed filling, endosperm development, and cell division in wheat [151]. Higher Z and ZR contents in seeds can improve seed filling rate in the early and middle stages of seed filling and are associated with seed filling rate in rice and maize [152, 153]. Zeatin and ZR increase simultaneously with the peak of endosperm mitotic activity during seed filling [154]. Exogenous application of GA in maize improved the degree of grain filling by increasing the levels of auxin, GA, Z and ABA in the grains [155]. It was also found that auxin, GA and Z content in grains was positively associated with grain mass and filling degree of grains. Cytokinin oxidase decreases cytokinins content in the later stages of seed development [58]. Cytokinin content is related to flower development, grain filling and endosperm growth in rice [152]. Grain seeds also have high cytokinin contents during endosperm development, and cytokinins promote cell division in the early stages of seed filling [152]. In addition, cytokinins and GA have antagonistic effects on various processes of seed development [156]. The levels of Z, ZR, ABA, and IAA in maize were positively correlated with seed filling rate and negatively correlated with the GA levels [157]. In wheat, the levels of Z, ZR, ABA, and IAA were positively correlated with seed filling rate and seed mass, but the ethylene content was negatively correlated [158, 159]. In maize, ABA, Z, and ZR contents were also positively related to seed mass and seed filling rate, but GA content was not. Seed filling rate was dramatically increased when ABA and Z content were higher and GA content was lower [153]. ABA, IAA and ZR contents in maize seeds increased dramatically during the early stages of seed filling and decreased gradually until maturity [160]. Similarly, Z and ZR contents of maize gradually increased during the early stages of seed filling, while the GA content decreased [153]. Moreover, the fluctuations in ZR and IAA contents were similar, they briefly increased in the early stages of grain filling and then decreased in the kernels [155]. The contents of IAA and Z + ZR affect the seed filling rate of maize and are normally located in the endosperm, as they are required for cell division [161]. IAA has been proposed as a correlative signal from seeds that regulates the development of other organs [162]. In maize seeds, high IAA and low ethylene content were significantly associated with grain filling rate [129] and high IAA content in seeds increased cytokinins content [163]. These observations were confirmed by Ahmad [164] who reported that IAA and cytokinin content play an important role in grain filling of maize at early stages by regulating endosperm cell division and thereby increasing seed filling rate. In soybean, IAA concentration and seed filling rate were independent [145]. Seed filling in maize appears to be dependent on IAA synthesis and cell wall invertase activity [128]. The absence of endogenous auxin in the embryo could be lethal [128], indicating the critical functions of the phytohormones in seed development and germination. Invertase activity, together with auxin transport, is important in regulating pathways of carbon cleavage during early development. Sugar signaling is thought to increase phloem transport and sugar import into endosperm cells via invertase activity [165]. Ethylene is also involved in cell division and grain filling [112]. Higher ethylene concentration leads to lower cell division, grain filling, and starch concentration, and higher ethylene concentration leads to higher soluble sugar content in growing rice endosperm [166]. Since cytokinin is known to regulate cell number and cell division activity of rice endosperm, the deleterious effects of ethylene on grain filling and cell division could be mediated by its antagonistic effect with cytokinin [166]. Apart from rice, there are studies reporting that the effect of ethylene is antagonistic to cytokinin [167], because ethylene production in plant tissues promotes cytokinin inactivation [168]. The ratio of ABA to ethylene regulates grain filling rate in wheat [112].

These studies show the importance of phytohormones during seed development. Therefore, seed filling is determined by the content and interactions of various plant hormones that regulate different metabolic processes related to the synthesis and accumulation of seed reserves [169].

5. Effect of environmental factors during seed filling

Stress in plants was described by Hans Selve in 1936 as "unfavorable conditions and environmental constraints in plants". This general definition can be applied to all organisms, but the definition of stress in plants differs from that in animals and humans. Plants are sedentary and live in fixed locations. Therefore, they cannot escape abiotic stress conditions when exposed to them and are constantly exposed to these conditions without protection. Animals, on the other hand, are mobile and can avoid and escape these conditions when needed. Since plants are sedentary, they need mechanisms to protect themselves from stressful conditions so that they can continue their vital activities [170].

Global warming and drought in the world have become important inhibitors of agricultural production in recent years. The process of seed filling, which is affected by environmental factors, is becoming increasingly important for agricultural production because the potential heat and drought affect the, and rate and duration of seed filling. To overcome these adverse conditions, plant breeders are developing new varieties that are resistant to biotic and abiotic stress conditions while ensuring efficient water and nutrient use and good yields. After pollination and fertilization, seed development begins with cell division for embryo and endosperm development in the ovule and continues with cell expansion and differentiation to form seeds. Seed formation continues with the accumulation of storage reserves such as carbohydrates, proteins, and lipids. After accumulation is complete, desiccation occurs, during which the seeds lose moisture.

Flowering plants reproduce by the production, dispersal and germination of seeds. The cellular stage includes all processes involved in the formation and development of the various parts of a seed. At this stage, the storage reserves for the embryo are synthesized and seed filling takes place. Many factors influence seed production and seed content. The position of seeds on the inflorescence can affect the duration and rate of seed filling. Seeds farthest from the transport source, such as seeds on a cob, may remain small because they do not receive sufficient nutrients for optimal seed growth. In the early stages of seed development, a constant and adequate supply of nutrients is required for seed production. Seeds that do not receive an adequate supply of nutrients during the generation stage may fail to develop or develop poorly and have a smaller seed mass. Plants can be affected by abiotic stress at any stage of development, but the generative stage is the most critical period when plants respond to stress conditions. Stress conditions during the generative stage adversely affect pollen formation, pollination and fertilization rates, and reduce fruiting and seed set, resulting in yield losses. The generative stage is highly susceptible to drought, cold, and heat, and these stress factors reduce fertilization, seed development, and the filling process [171]. Heat stress has significant negative effects on meiosis during pollen development and could greatly reduce pollen fertility, pollen quantity and quality, pollen germination, and pollen tube development on the stigma [172, 173]. Heat stress can also significantly affect seed development during the seed filling stage due to reduced assimilate supply and reduce seed yield in many crops including cereals and legumes [174, 175]. The seed filling period is also closely related to the development of plant senescence [129]. Heat and drought stress during the seed filling period cause early senescence and also shorten the seed filling time [176, 177].

5.1 Effects of drought stress during seed filling

Drought stress limits vegetative growth by reducing leaf water content and stomatal conductance [178] in various crops such as cereals [179, 180] and legumes [181]. Decreased stomatal conductance increases leaf temperature, and both events lead to wilt symptoms [182, 183]. Drought stress damages cell membranes [184, 185], decreases chlorophyll content, photosynthesis, and reserve synthesis [178, 186–188]. Drought stress also impairs plant nutrient uptake [189, 190] and significantly reduces nitrogen fixation in legumes such as soybean [191] and pea [192]. The overall negative effects of drought stress reduce the production of assimilates and reduce the transport of reserves to the developing seeds of plants [193–195].

The generative phase of plants is more sensitive to drought stress than the vegetative phase. Drought stress reduces the number of flowers, fruits, and seed set and therefore could reduce seed yield [196, 197]. Decreased water content in tissues leads to a reduction in the activity of the acid invertase enzyme, which in turn prevents sucrose uptake into developing seeds [198]. Low sucrose and high ABA levels lead to poor seed development in cereals under drought stress [199]. Wheat plants subjected to drought stress during the seed filling period showed a significant decrease in cell wall and soluble invertase activities, and glucose, fructose, and sucrose contents of the drought-sensitive genotype were significantly lower [200]. Drought in the early stages of seed development leads to a reduction in seed size due to reduced number of endosperm cells. Seed yield was significantly reduced in plants subjected to drought stress during the seed filling stage [186, 201, 202]. Drought stress in the early stages of seed filling reduced germination percentage in soybean by up to 9% compared to the control [203]. Similarly, in chickpea, medium-sized seeds produced under drought stress had lower germination percentage and viability than control seeds [204].

Drought stress during embryogenesis and seed filling reduces the number of endosperm cells formed and thus the size and weight of the seeds [205]. At this stage, the duration and amount of seed storage reserves, such as starch accumulation in the endosperm decrease, and so does seed weight [206]. Drought stress during seed filling reduces the number and size of starch granules in endosperm cells [206]. Drought stress affects the composition of seed reserves. The starch content of wheat seeds subjected to drought stress during seed filling is significantly reduced [207]. Drought stress negatively affects phthosynthesis, and low phthosynthesis product content inhibits starch biosynthesis [208] and related activities such as reduced endosperm cell number and starch granule size [206] and lower starch amylase content [209].

Lipid content and fatty acid composition change due to lower content of soluble sugars such as glucose, fructose, and sucrose and reduced transport of sugars from the phloem to endosperm cells under drought conditions [210]. In peanuts, the content of linoleic and behenic acids decreased, while the content of stearic and oleic acids increased under drought stress [211]. In maize, drought stress resulted in a significant decrease in oil content, while the content of linolenic and oleic acids in the seeds increased. In addition to the oil content, the total tocopherol, flavonoids, and oil-soluble phenolics contents also decreased [212, 213]. In soybean, drought stress decreased the oil content and oleic acid content of seeds [214]. While drought stress reduced the starch and oil content of seeds, the protein content of soybean seeds grown under drought stress increased. Seed nitrogen supply depends on remobilization of nitrogen from vegetative tissues, while starch and oil biosynthesis depends on sugars from photosynthesis, which decreases under drought stress. For this reason, seed viability may also be affected by drought during late maturation and seed desiccation.

5.2 Effects of heat stress during seed filling

Heat stress affects all stages of plant development from germination to senescence. Different plants have different sensitivities to heat stress during seed filling [215, 216].

Seed filling rate and potential seed mass are generally used as two selection criteria for heat stress tolerance [217]. High temperature stress could accelerate seed filling rate by shortening the duration of seed filling and could lead to yield reductions [216, 218]. Heat stress during seed filling significantly reduces seed weight, seed number and seed yield in legumes [219–221], cereals [222], and other crops [223]. In chickpea [224] and lentil [183] increased seed filling rate resulted in smaller seed size. Similarly, a reduction in seed filling time resulted in smaller seed sizes in soybean, pea, and white lupin [225]. Temperature also affects seed filling rate and duration. An increase in ambient temperature from 15.5°C to 26.6°C decreased seed filling duration in cowpea from 21 days to 14 days [226]. A 1.7°C increase in temperature shortened the duration of seed filling and accelerated maturation, but decreased seed yield in chickpea [227].

Starch accounts for more than 65% of the dry weight of cereal seeds [228]. Therefore, the main reason for yield reduction is mainly the reduction in starch accumulation. Heat stress during the seed filling period reduces seed size and mass in wheat [229] and rice [230], and also impedes starch biosynthesis and accumulation by altering the expression of genes in starch biosynthetic pathways [231]. As a result of altered gene expression, the amount of non-structural carbohydrates decrease, altering the balance between soluble sugars and starch [232]. Heat stress decreases the content of sugars such as fructose and hexose phosphate in wheat [233]. In some cases, up-regulation of starch hydrolyzing enzymes such as α-amilaz under heat stress is thought to be responsible for the increased sugar content during seed filling [234, 235]. Thus, heat stress negatively affects starch accumulation by altering gene expressions in metabolic pathways. These changes may vary depending on the duration of heat stress, the growing season, and the plant species.

Oil content and quality are severely affected by heat stress in oliferous crops [236]. The effects of heat stress may vary depending on location, altitude, precipitation, and differences between day and night temperatures during the seed filling period. Because oleic acid and linoleic acid are produced by the same pathway through desaturation, there is a strong and negative correlation between them, and temperature and precipitation during the flowering and seed filling periods have significant effects on the fatty acid composition of plants [237]. Growth experiments conducted at different temperatures (10, 16, 21, 26.5°C) with canola, flax, sunflower, safflower and castor bean during the seed filling period have shown that the fatty acid composition changes and the amount of unsaturated fatty acids is reduced, with the exception of safflower and castor bean [238, 239]. High temperatures during seed filling reduce linoleic acid content and increase oleic acid content in seeds, while palmitic and stearic acid content change insignificantly [240–242]. The fatty acid composition of rapeseed also changes depending on location and year. While low temperatures and precipitation decreased oleic acid content, high temperatures and low precipitation did not cause significant changes in linoleic and linolenic acid content [243]. The activities of oleayl-PC desaturase and linoleayl-PC desaturase, which catalyze the synthesis of linoleic and linolenic acids from oleic acid, are decreased by high temperatures [244]. Consequently, high temperatures have negative effects on linoleic and linolenic acids synthesis, whereas high temperatures have positive effects on oleic acid synthesis [240, 241, 245]. Linolenic acid is the major fatty acid in flaxseed, and increasing temperatures (15, 20, 25, 30°C) during seed filling reduced the oleic and linolenic acids content in flaxseed [246]. Increasing the growth temperature from 29°C to 35°C during seed filling in sunflower and soybean resulted in a 2.6% reduction in oil content in the seeds of these plants [214, 247].

Heat stress reduces the duration of seed filling, the amount of protein accumulation and protein quality, but has no effect on the rate of accumulation [248]. The composition and quality of storage proteins change due to changes in nitrogen mobilization during seed filling in wheat under heat stress [228, 249]. A decrease in high molecular weight glutenins and increased gliadin accumulation decreased dough quality in wheat under heat stress [248]. Similarly, high temperatures caused denaturation and aggregation of several storage proteins (globulins, legumin, and vicilin) in pea [250] and loss of enzyme activities in protein synthesis in lupin [251].

6. Conclusion

Effects of climate change and the growing world population along with reduced natural resources pose a major challenge for crop production and food security, especially in the developing countries. In this chapter, we have provided an overview of genetic regulation and synthesis of seed reserves, the role of phytohormones, and the influence of environmental factors on metabolic processes during seed filling in different plant species. To increase crop productivity, it is necessary to understand seed development processes together with reserve synthesis and accumulation under normal and stress conditions, and the control and regulation at the molecular and hormonal levels in seeds during seed filling. We have summarized recent studies on seed filling processes by highlighting the effects of phytohormones on the seed filling process and their interactions the effects of abiotic stress conditions on seed reserve quality and yield during the seed filling process.

7. Future perspectives

Future work should be directed toward investigating various signaling and metabolic pathways during seed filling, and developing a feasible system for delimitating roles of different genes, their regulation and interactions during seed development to better understand roles of these reserve metabolites and their interactions. In addition, future studies could process RNAs, proteins, and metabolites as quantitative traits, offering new insights into the integration of omics tools and how these traits may regulate seed filling as well. In the future, these new insights will help to understand the biochemistry and physiology of the seed filling process at the molecular level and manipulate metabolic pathways to improve valuable seed traits through metabolic engineering. The duration of seed filling is under genetic control and influenced by environmental factors. Therefore, it could be used as a selection criterion in plant breeding programs that focus on yield enhancement. There is a need for multidisciplinary studies to delimitate steps in the synthetic pathways leading to seed storage compounds, and their regulation. In recent years, progress has been made in understanding the various aspects of seed development including seed filling, maturation, acquisition of desiccation tolerance, and post-maturation stages. Identification of important genes and regulatory pathways related to seed filling processes will provide useful tools for developing better strategies to improve seed production. New omics studies will expand our understanding of the processes associated with seed filling. Modeling the stages associated with seed filling, and seed quality will provide insights into seed development and lead to improved seed yield and quality.

Acknowledgements

Sercan Önder and Murat Mutlucan were financially supported by the Council for Higher Education under the YÖK 100/2000 scholarship program for graduate students.

Conflict of interest

The authors declare no conflict of interest.

Author details

Sercan Önder[1*], Sabri Erbaş[2], Damla Önder[3], Muhammet Tonguç[1] and Murat Mutlucan[2]

1 Department of Agricultural Biotechnology, Faculty of Agriculture, Isparta University of Applied Sciences, Isparta, Turkey

2 Department of Field Crops, Faculty of Agriculture, Isparta University of Applied Sciences, Isparta, Turkey

3 Department of Biology, Faculty of Arts and Sciences, Suleyman Demirel University, Isparta, Turkey

*Address all correspondence to: sercanonder@isparta.edu.tr

References

[1] Bewley JD, Bradford KJ, Hilhorst HWM, Nonogaki H. Seeds: Physiology of Development, Germination and Dormancy. 3th ed. Heidelberg Dordrecht London, USA: Springer, New York; 2013. ISBN 978-1-4614-4692-7

[2] Alamillo JM, Diaz-Leal JL, Sanchez-Moran MV, Pineda M. Molecular analysis of ureide accumulation under drought stress in *Phaseolus vulgaris* L. Plant, Cell Environment. 2010;**33**:1828-1837

[3] Desimone M, Catoni E, Ludewig U, Hilpert M, Schneider A, Kunze R, et al. A novel superfamily of transporters for allantoin and other oxo derivatives of nitrogen heterocyclic compounds in Arabidopsis. The Plant Cell. 2002;**14**:847-856

[4] Hanks JF, Tolbert NE, Schubert KR. Localization of enzymes of ureide biosynthesis in peroxisomes and microsomes of nodules. Plant Physiology. 1981;**68**:65-69

[5] Lin Y, Sun L, Nguyen LV, Rachubinski RA, Goodman HM. The pex16b homologue SSE1 and storage organelle formation in Arabidopsis seeds. Science. 1999;**284**:328-330

[6] Wobus U, Weber H. Sugars as signalmolecules in plant seed development. Biological Chemistry. 1999;**380**:937-944

[7] Rolletschek H, Koch K, Wobus U, Borisjuk L. Positional cues for the starch/lipid balance in maize kernels and resource partitioning to the embryo. Plant Journal. 2005;**42**:69-83

[8] Dale EM, Housely TL. Sucrose synthase activity in developing wheat endosperms differing in maximum weight. Plant Physiology. 1986;**82**:7-10

[9] Briarty LG, Hughes CE, Evers AD. The developing endosperm of wheat—A stereological analysis. Annals of Botany. 1979;**44**:641-658

[10] Olsen OA. Endosperm development: Cellularization and cell fate specification. Annual Review of Plant Physiology and Plant Molecular Biology. 2001;**52**:233-267

[11] Brocklehurst PA. Factors controlling grain weight in wheat. Nature. 1977;**266**:348-349

[12] Tetlow IJ. Starch biosynthesis in developing seeds. Seed Science Research. 2011;**21**:532. DOI: 10.1017/S0960258510000292

[13] Bihmidine S, Hunter CT, Johns CE, Koch KE, Braun DM. Regulation of assimilate import into sink organs: Update on molecular drivers of sink strength. Frontiers in Plant Science. 2013;**4**:1-15. DOI: 10.3389/fpls.2013.00177

[14] Greenberg E, Preiss J. The occurrence of adenosine diphosphate glucose:Glycogen transglucosylase in bacteria. Journal of Biological Chemistry. 1964;**239**:4314-4315

[15] Parry MAJ, Hawkesford MJ. Food security: Increasing yield and improving resource use efficiency. Proceedings of the Nutrition Society. 2010;**69**:592-600. DOI: 10.1017/S0029665110003836

[16] Gross P, Ap Rees T. Alkaline inorganic pyrophosphatase and starch synthesis in amyloplasts. Planta. 1986;**167**:140-145. DOI: 10.1007/BF00446381

[17] Weber APM, Schwacke R, Flügge U-I. Solute transporters of the plastid envelope membrane. Annual Review Plant Biology. 2005;**56**:133-164. DOI: 10.1146/annurev.arplant.56.032604.144228

[18] Preiss J, Sivak M. Starch synthesis in sinks and sources. In: Zamski E, Schaffter AA, editors. Photoassimilate Distribution in Plants and Crops. New York: Marcel Dekker; 1996. pp. 63-69

[19] Denyer K, Dunlap F, Thorbjørnsen T, Keeling P, Smith AM. The major form of ADPglucose pyrophosphorylase in maize endosperm is extraplastidial. Plant Physiology. 1996;**112**:779-783

[20] Thorbjørnsen T, Villand P, Denyer K, Olsen OA, Smith AM. Distinct isoforms of ADPglucose pyrophosphorylase occur inside and outside the amyloplasts in barley endosperm. Plant Journal. 1996;**10**:243-250

[21] Sikka VK, Choi S, Kavakli IH, Sakulsingharoj C, Gupta S, Ito H, et al. Subcellular compartmentation and allosteric regulation of the rice endosperm ADPglucose pyrophosphorylase. Plant Science. 2001;**161**:461-468

[22] Tetlow IJ, Davies EJ, Vardy KA, Bowsher CG, Burrell MM, Emes MJ. Subcellular localization of ADPglucose pyrophosphorylase in developing wheat endosperm and analysis of a plastidial isoform. Journal of Experimental Botany. 2003;**54**:715-725

[23] Rees TA, Morrell S. Carbohydrate metabolism in developing potatoes. American Potato Journal. 1990;**6**:835-847

[24] Kleczkowski LA. Glucose activation and metabolism through UDP-glucose pyrophosphorylase in plants. Phytochemistry. 1994;**37**:1507-1515

[25] Tetlow IJ, Bowsher CG, Scrase-Field EFAL, Davies EJ, Emes MJ. The synthesis and transport of ADPglucose in cereal endosperms. Journal of Applied Glycoscience. 2003;**50**:231-236

[26] Bowsher CG, Scrase-Field EFAL, Esposito S, Emes MJ, Tetlow IJ. Characterization of ADP-glucose transport across the cereal endosperm amyloplast envelope. Journal of Experimental Botany. 2007;**58**:1321-1332

[27] Nakamura Y, Kawaguchi K. Multiple forms of ADP-glucose pyrophosphorylase of rice endosperm. Physiologia Plantarum. 1992;**84**:336-342

[28] Villand P, Aalen R, Olsen OA, Lonneborg A, Luthi E, Kleczkowski LA. PCR-amplification and sequence of cDNA clones for the small and large subunits of ADP-glucose pyrophosphorylase from barley tissues. Plant Molecular Biology. 1992;**19**:381-389

[29] La Cognata U, Willmitzer L, Muller-Rober B. Molecular cloning and characterisation of novel isoforms of potato ADP-glucose pyrophosphorylase. Molecular and General Genetics. 1995;**246**:538-548

[30] Villand P, Olsen OA, Killan A, Kleczkowski LA. ADPglucose pyrophosphorylase large subunit cDNA from barley endosperm. Plant Physiology. 1992;**100**:1617-1618

[31] Olive MR, Ellis RJ, Schuch WW. Isolation and nucleotide sequences of cDNA clones encoding ADPglucose pyrophosphorylase polypeptides from wheat leaf and endosperm. Plant Molecular Biology. 1989;**12**:525-538

[32] Lee SK, Hwang SK, Han M, Eom JS, Kang HG, Han Y, et al. Identification of the ADP-glucose pyrophosphorylase

isoforms essential for starch synthesis in the leaf and seed endosperm of rice (*Oryza sativa* L.). Plant Molecular Biology. 2007;**65**:531-546

[33] Muller-Rober B, Kossmann J, Hannah LC, Willmitzer L, Sonnewald U. Only one of two different ADPglucose pyrophosphorylase genes from potato responds strongly to elevated levels of sucrose. Molecular and General Genetics. 1990;**224**:136-146

[34] Duwenig E, Steup M, Kossmann J. Induction of genes encoding plastidic phosphorylase from spinach (*Spinacia oleracea* L.) and potato (*Solanum tuberosum* L.) by exogenously supplied carbohydrates in excised leaf discs. Planta. 1997;**203**:111-120

[35] De Fekete MAR, Leloir LF, Cardini CE. Mechanism of starch biosynthesis. Nature. 1960;**187**:918-919

[36] Nelson OE, Rines HW. The enzymatic deficiency in the waxy mutant of maize. Biochemical and Biophysical Research Communications. 1962;**9**:297-300

[37] Hanashiro I, Itoh K, Kuratomi Y, Yamazaki M, Igarishi T, Matsugasako J, et al. Granule-bound starch synthase I is responsible for biosynthesis of extralong chains of amylopectin in rice. Plant and Cell Physiology. 2008;**49**:925-933

[38] Ball SG, Morell MK. From bacterial glycogen to starch: Understanding the biogenesis of the plant starch granule. Annual Review of Plant Biology. 2003;**54**:207-233

[39] Tunail N. Chemical microbiology. Ankara: Ankara University Agriculture of Faculty Press; 1975. p. 86

[40] Baydar H, Erbaş S. Oil crops sciences and technology. Isparta: Suleyman Demirel University Press, Press No: 97; 2014. p. 313

[41] Browse J, Somerville C. Glycerolipid synthesis: Biochemistry and regulation. Annual Review of Plant Physiology and Plant Molecular Biology. 1991;**42**:467-506

[42] Voelker TA, Worrell AC, Anderson L, Bleibaum J, Fan C, Hawkins DJ, et al. Fatty acid biosynthesis redirected to medium chains in transgenic oilseed plants. Science. 1992;**257**:72-74

[43] Weselake RJ, Taylor DC, Rahman MH, Shah S, Laroche A, McVetty PBE, et al. Increasing the flow of carbon into seed oil. Biotechnology Advances. 2009;**27**:866-878

[44] Bates PD, Stymne S, Ohlrogge J. Biochemical pathways in seed oil synthesis. Current Opinion in Plant Biology. 2013;**16**:358-364

[45] Shewry PR, Tatham AS. The prolamin storage proteins of cereal seeds: Structure and evolution (review article). Biochemical Journal. 1990;**267**:1-12

[46] Vitale A, Ceriotti A, Denecke J. The role of the endoplasmic reticulum in protein synthesis, modification and intracellular transport. Journal of Experimental Botany. 1993;**44**:1417-1444

[47] Pelham HRB. Control of protein exit from the endoplasmic reticulum. Annual Review of Cell Biology. 1989;**5**:1-23

[48] Chrispeels ML. Sorting of proteins in the secretory system. Annual Review of Plant Physiology and Plant Molecular Biology. 1991;**42**:21-53

[49] Kermode AR. Mechanisms of intracellular protein transport and targeting in plant cells. Critical Reviews in Plant Sciences. 1996;**15**:285-423

[50] Blond-Elguindi S, Cwirla SE, Dower WJ, Lipshutz RJ, Sprang SR, Sambrook F, et al. Affinity panning of a library of peplides displayed on bacteriophages reveals the binding specificity of BiP. Cell. 1993;**75**:717-772

[51] Muntz K. Proteases and proteolytic cleavage of storage proteins in developing and germinating dicotyledpnous seeds. Journal of Experimental Botany. 1996;**47**:605-622

[52] Chrispeels MJ, Tague BW. Transport and targeting of proteins to protein storage vacuoles (protein bodies) in developing seeds. In: Taylorson RB, editor. Recent Advances ill Development and Germination o/Seedv. New York: Plenum Pub!. Corp.; 1990

[53] Gething ML, Sambrook F. Protein folding in the cell. Nature. 1992;**355**:33-45

[54] Sun X, Shantharaj D, Kang X, Ni M. Transcriptional and hormonal signaling control of Arabidopsis seed development. Current Opinion in Plant Biology. 2010;**13**:611-620. DOI: 10.1016/j.pbi.2010.08.009

[55] Gao MJ, Gropp G, Wei S, Hegedus DD, Lydiate DJ. Combinatorial networks regulating seed development and seed filling. In: Ken-ichi Sato editor. Embryogenesis. 1st ed. InTech Rijeka; 2012. DOI: 10.5772/35960

[56] Weber H, Borisjuk L, Wobus U. Molecular physiology of legume seed development. Annual Review of Plant Biology. 2005;**56**:253-279. DOI: 10.1146/annurev.arplant.56.032604.144201

[57] North H, Baud S, Debeaujon I, Dubos C, Dubreucq B, Grappin P, et al. Arabidopsis seed secrets unravelled after a decade of genetic and omicsdriven research. Plant Journal. 2010;**61**:971-981. DOI: 10.1111/j.1365-313X.2009.04095.x

[58] Agrawal GK, Rakwal R. Seed development: OMICS technologies toward improvement of seed quality and crop yield. 1st ed. Dordrecht: Springer; 2012. 576 p. DOI: 10.1007/978-94-007-4749-4

[59] Yin S, Li P, Xu Y, Liu J, Yang T, Wei J, et al. Genetic and genomic analysis of the seed-filling process in maize based on a logistic model. Heredity. 2020;**124**:122-134. DOI: 10.1038/s41437-019-0251-x

[60] Holdsworth MJ, Bentsink L, Soppe WJ. Molecular networks regulating Arabidopsis seed maturation, after-ripening, dormancy and germination. New Phytologist. 2008;**179**:33-54. DOI: 10.1111/j.1469-8137.2008.02437.x

[61] Meitzel T, Radchuk R, McAdam EL, Thormählen I, Feil R, Munz E, et al. Trehalose 6-phosphate promotes seed filling by activating auxin biosynthesis. New Phytologist. 2021;**229**:1553-1565. DOI: 10.1111/nph.16956

[62] Chen B, Zhang G, Li P, Yang J, Guo L, Benning C, et al. Multiple GmWRI1s are redundantly involved in seed filling and nodulation by regulating plastidic glycolysis, lipid biosynthesis and hormone signalling in soybean (*Glycine max*). Plant Biotechnology Journal. 2020;**18**:155-171. DOI: 10.1111/pbi.13183

[63] Yang T, Guo L, Ji C, Wang H, Wang J, Zheng X, et al. The B3 domain-containing transcription factor ZmABI19 coordinates expression of key factors required for maize seed development and grain filling. The Plant Cell. 2021;**33**:104-128. DOI: 10.1093/plcell/koaa008

[64] Mathew IE, Priyadarshini R, Mahto A, Jaiswal P, Parida SK, Agarwal P. SUPER STARCHY1/ONAC025 participates in rice grain filling. Plant Direct. 2020;**4**:e00249. DOI: 10.1002/pld3.249

[65] Shen L, Luo G, Song Y, Xu J, Ji J, Zhang C, et al. A novel NAC family transcription factor SPR suppresses seed storage protein synthesis in wheat. Plant Biotechnology Journal. 2020;**19**:992. DOI: 100710.1111/pbi.13524

[66] To A, Valon C, Savino G, Guilleminot J, Devic M, Giraudat J, et al. A network of local and redundant gene regulation governs *Arabidopsis* seed maturation. The Plant Cell. 2006;**18**:1642-1651. DOI: 10.1105/tpc.105.039925

[67] Grossniklaus U, Schneitz K. The molecular and genetic basis of ovule and megagametophyte development. Seminars in Cell & Developmental Biology. 1998;**9**:227-238. DOI: 10.1006/scdb.1997.0214

[68] Tomohiro K, Ohad N, Yadegari R, Hannon M, Dinneny J, Wells D, et al. Control of fertilization-independent endosperm development by the MEDEA polycomb gene in Arabidopsis. Proceedings of the National Academy of Sciences. 1999;**96**:4186-4191. DOI: 10.1073/pnas.96.7.4186

[69] Ohad NIR, Margossian L, Hsu YC, Williams C, Repetti P, Fischer RL. A mutation that allows endosperm development without fertilization. Proceedings of the National Academy of Sciences. 1996;**93**:5319-5324. DOI: 10.1073/pnas.93.11.5319

[70] Kohler C, Hennig L, Spillane C, Pien S, Gruissem W, Grossniklaus U. The Polycomb-group protein MEDEA regulates seed development by controlling expression of the MADS-box gene PHERES1. Genes and Development. 2003;**17**:1540-1553. DOI: 10.1101/gad.257403

[71] Ingouff M, Haseloff J, Berger F. Polycomb group genes control developmental timing of endosperm. The Plant Journal. 2005;**42**:663-674. DOI: 10.1111/j.1365-313X.2005.02404.x

[72] Makarevich G, Leroy O, Akinci U, Schubert D, Clarenz O, et al. Different Polycomb group complexes regulate common target genes in Arabidopsis. EMBO Reports. 2006;**7**:947-952. DOI: 10.1038/sj.embor.7400760

[73] Tiwari S, Schulz R, Ikeda Y, Dytham L, Bravo J, Mathers L, et al. Maternally expressed PAB C-Terminal, a novel imprinted gene in Arabidopsis, encodes the conserved C-terminal domain of polyadenylate binding proteins. The Plant Cell. 2008;**20**:2387-2398. DOI: 10.1105/tpc.108.061929

[74] Kinoshita T, Miura A, Choi Y, Kinoshita Y, Cao X, et al. One-way control of FWA imprinting in Arabidopsis endosperm by DNA methylation. Science. 2004;**303**:521-523. DOI: 10.1126/science.1089835

[75] Köhler C, Hennig L. Regulation of cell identity by plant Polycomb and trithorax group proteins. Current Opinion in Genetics & Development. 2010;**20**:541-547. DOI: 10.1016/j.gde.2010.04.015

[76] Finnegan EJ, Dennis ES. Isolation and identification by sequence homology of a putative cytosine methyltransferase from *Arabidopsis thaliana*. Nucleic Acids Research. 1993;**21**:2383-2388. DOI: 10.1093/nar/21.10.2383

[77] Kankel MW, Ramsey DE, Stokes TL, Flowers SK, Haag JR, Jeddeloh JA, et al. *Arabidopsis* MET1 cytosine methyltransferase mutants. Genetics. 2003;**163**:1109-1122. DOI: 10.1093/genetics/163.3.1109

[78] FitzGerald J, Luo M, Chaudhury A, Berger F. DNA methylation causes

predominant maternal controls of plant embryo growth. PLoS One. 2008;**3**:e2298. DOI: 10.1371/journal.pone.0002298

[79] Drea S, Leader DJ, Arnold BC, Shaw P, Dolan L, Doonan JH. Systematic spatial analysis of gene expression during wheat caryopsis development. The Plant Cell. 2005;**17**:2172-2185. DOI: 10.1105/tpc.105.034058

[80] Sreenivasulu N, Altschmied L, Radchuk V, Gubatz S, Wobus U, Weschke W. Transcript profiles and deduced changes of metabolic pathways in maternal and filial tissues of developing barley grains. The Plant Journal. 2004;**37**:539-553. DOI: 10.1046/j.1365-313X.2003.01981.x

[81] Zhu T, Budworth P, Chen W, Provart N, Chang HS, Guimil S, et al. Transcriptional control of nutrient partitioning during rice grain filling. Plant Biotechnology. 2003;**1**:59-70. DOI: 10.1046/j.1467-7652.2003.00006

[82] Verdier J, Kakar K, Gallardo K, Le Signor C, Aubert G, Schlereth A, et al. Gene expression profiling of *M. truncatula* transcription factors identifies putative regulators of grain legume seed filling. Plant Molecular Biology. 2008;**67**:567-580. DOI: 10.1007/s11103-008-9320-x

[83] Jin X, Fu Z, Lv P, Peng Q, Ding D, Li W, et al. Identification and characterization of microRNAs during maize grain filling. PLoS One. 2015;**10**:e0125800. DOI: 10.1371/journal.pone.0125800

[84] Peng T, Sun H, Du Y, Zhang J, Li J, Liu Y, et al. Characterization and expression patterns of microRNAs involved in rice grain filling. PLoS One. 2013;**8**:e54148. DOI: 10.1371/journal.pone.0054148

[85] Yi R, Zhu Z, Hu J, Qian Q, Dai J, Ding Y. Identification and expression analysis of microRNAs at the grain filling stage in rice (*Oryza sativa* L.) via deep sequencing. PLoS One. 2013;**8**:e57863. DOI: 10.1371/journal.pone.0057863

[86] Wang S, Wu K, Yua Q, Liu X, Liu Z, Lin X, et al. Control of grain size, shape and quality by OsSPL16 in rice. Nature Genetics. 2012;**44**:950-954. DOI: 10.1038/ng.2327

[87] Zhang YC, Yu Y, Wang CY, Li ZY, Liu Q, Xu J, et al. Overexpression of microRNA OsmiR397 improves rice yield by increasing grain size and promoting panicle branching. Nature Biotechnology. 2013;**3**:848-852. DOI: 10.1038/nbt.2646

[88] Baud S, Mendoza MS, To A, Harscoët E, Lepiniec L, Dubreucq B. WRINKLED1 specifies the regulatory action of LEAFY COTYLEDON2 towards fatty acid metabolism during seed maturation in *Arabidopsis*. Plant Journal. 2007;**50**:825-838. DOI: 10.1111/j.1365-313X.2007.03092.x

[89] Santos-Mendoza M, Dubreucq B, Miquel M, Caboche M, Lepiniec L. LEAFY COTYLEDON 2 activation is sufficient to trigger the accumulation of oil and seed specific mRNAs in *Arabidopsis* leaves. FEBS Letter. 2005;**579**:4666-4670. DOI: 10.1016/j.febslet.2005.07.037

[90] Kagaya Y, Okuda R, Ban A, Toyoshima R, Tsutsumida K, Usui H, et al. Indirect ABA-dependent regulation of seed storage protein genes by FUSCA3 transcription factor in Arabidopsis. Plant Cell Physiology. 2005;**46**:300-311. DOI: 10.1093/pcp/pci031

[91] Rook F, Hadingham SA, Li Y, Bevan MW. Sugar and ABA response pathways and the control of gene expression. Plant, Cell & Environment.

2006;**29**:426-434. DOI: 10.1111/j.1365-3040.2005.01477.x

[92] Rolland F, Baena-Gonzalez E, Sheen J. Sugar sensing and signaling in plants: Conserved and novel mechanisms. Annual Review of Plant Biology. 2006;**57**:675-709. DOI: 10.1146/annurev.arplant.57.032905.105441

[93] Tegeder M, Wang XD, Frommer WB, Offler CE, Patrick JW. Sucrose transport into developing seeds of *Pisum sativum* L. The Plant Journal. 1999;**18**:151-161. DOI: 10.1046/j.1365-313X.1999.00439.x

[94] Weschke W, Panitz R, Sauer N, Wang Q, Neubohn B, Weber H, et al. Sucrose transport into barley seeds: Molecular characterization of two transporters and implications for seed development and starch accumulation. The Plant Journal. 2000;**21**:455-467. DOI: 10.1046/j.1365-313x.2000.00695.x

[95] Tsukagoshi H, Morikami A, Nakamura K. Two B3 domain transcriptional repressors prevent sugar-inducible expression of seed maturation genes in *Arabidopsis* seedlings. Proceedings of the National Academy of Sciences of USA. 2007;**104**:2543-2547. DOI: 10.1073/pnas.0607940104

[96] Fincher GB. Molecular and cellular biology associated with endosperm mobilization in geminating cereal grain. Annual Review of Plant Physiology and Plant Molecular Biology. 1989;**40**:305-346. DOI: 10.1146/annurev.pp.40.060189.001513

[97] Baud S, Wuillème S, Lemoine R, Kronenberger J, Caboche M, Lepiniec L, et al. The AtSUC5 sucrose transporter specifically expressed in the endosperm is involved in early seed development in *Arabidopsis*. Plant Journal. 2005;**43**:824-836. DOI: 10.1111/j.1365-313X.2005.02496.x

[98] Ohdan T, Francisco PB, Sawada T, Hirose T, Terao T, Satoh H, et al. Expression profiling of genes involved in starch synthesis in sink and source organs of rice. Journal of Experimental Botany. 2005;**56**:3229-3244. DOI: 10.1093/jxb/eri292

[99] Smidansky ED, Meyer FD, Blakeslee B, Weglarz TE, Greene TW, Giroux MJ. Expression of a modified ADP-glucose pyrophosphorylase large subunit in wheat seeds stimulates photosynthesis and carbon metabolism. Planta. 2007;**225**:965-976. DOI: 10.1007/s00425-006-0400-3

[100] Hannah LC, Futch B, Bing J, Shaw JR, Boehlein S, Stewart JD, et al. A shrunken-2 transgene increases maize yield by acting in maternal tissues to increase the frequency of seed development. The Plant Cell. 2012;**24**:2352-2363. DOI: 10.1105/tpc.112.100602

[101] Tuncel A, Okita TW. Improving starch yield in cereals by overexpression of ADP glucose pyrophosphorylase: Expectations and unanticipated outcomes. Plant Science. 2013;**211**:52-60. DOI: 10.1016/j.plantsci.2013.06.009

[102] Gómez LD, Baud S, Gilday A, Li Y, Graham IA. Delayed embryo development in the *ARABIDOPSIS TREHALOSE-6-PHOSPHATE SYNTHASE 1* mutant is associated with altered cell wall structure, decreased cell division and starch accumulation. Plant Journal. 2006;**46**:69-84. DOI: 10.1111/j.1365-313X.2006.02662.x

[103] Fait A, Angelovici R, Less H, Ohad I, Urbanczyk-Wochniak E, Fernie AR, et al. *Arabidopsis* seed development and germination is associated with temporally distinct metabolic switches. Plant Physiology. 2006;**142**:839-854. DOI: 10.1104/pp.106.086694

[104] Canas RA, Quillere I, Christ A, Hirel B. Nitrogen metabolism in the developing ear of maize (*Zea mays*): Analysis of two lines contrasting in their mode of nitrogen management. New Phytologist. 2009;**184**:340-352. DOI: 10.1111/j.1469-8137.2009.02966.x

[105] Kang Y, Li M, Sinharoy S, Verdier J. A snapshot of functional genetic studies in *Medicago truncatula*. In: Bruijn F, editor. The Model Legume *Medicago truncatula*. 1st ed. Wiley; 2019. p. 7-30. DOI: 10.1002/9781119409144.ch02

[106] Gimeno-Gilles C, Gervais ML, Planchet E, Satour P, Limami AM, Lelievre E. A stress-associated protein containing A20/AN1 zing-finger domains expressed in *Medicago truncatula* seeds. Plant Physiology and Biochemistry. 2011;**49**:303-310. DOI: 10.1016/j.plaphy.2011.01.004

[107] van Der Geest A, Frisch DA, Kemp JD, Hall TC. Cell ablation reveals that expression from the phaseolin promoter is confined to embryogenesis and microsporogenesis. Plant Physiology. 1995;**109**:1151-1158. DOI: 10.1104/pp.109.4.1151

[108] Kadosh D, Struhl K. Targeted recruitment of the Sin3-Rpd3 histone deacetylase complex generates a highly localized domain of repressed chromatin in vivo. Molecular and Cellular Biology. 1998;**18**:5121-5127. DOI: 10.1128/MCB.18.9.5121

[109] Nambara E, Marion-Poll A. ABA action and interactions in seeds. Trends in Plant Science. 2003;**8**:213-217. DOI: 10.1016/S1360-1385(03)00060-8

[110] Okamoto M, Kuwahara A, Seo M, Kushiro T, Asami T, Hirai N, et al. CYP707A1 and CYP707A2, which encode abscisic acid 8 ¢- hydroxylases, are indispensable for proper control of seed dormancy and germination in *Arabidopsis*. Plant Physiology. 2006;**141**:97-107. DOI: 10.1104/pp.106.079475

[111] Millar AA, Jacobsen JV, Ross JJ, Helliwell CA, Poole AT, Scofield G, et al. Seed dormancy and ABA metabolism in Arabidopsis and barley: The role of ABA 8′-hydroxylase. Plant Journal. 2006;**45**:942-954. DOI: 10.1111/j.1365-313X.2006.02659.x

[112] Yang J, Zhang J, Liu K, Wang Z, Liu L. Abscisic acid and ethylene interact in wheat grains in response to soil drying during grain filling. New Phytologist. 2006;**171**:293-303. DOI: 10.1111/j.1469-8137.2006.01753.x

[113] Nguyen HN, Sabelli PA, Larkins BA. Endoreduplication and programmed cell death in the cereal endosperm. In: Olsen OA, editor. Endosperm: Development and Molecular Biology. 8th ed. Berlin: Springer; 2007. pp. 21-43. DOI: 10.1007/7089_2007_107

[114] Seo M, Hanada A, Kuwahara A, Endo A, Okamoto M, Yamauchi Y, et al. Regulation of hormone metabolism in Arabidopsis seeds: Phytochrome regulation of abscisic acid metabolism and abscisic acid regulation of gibberellin metabolism. Plant Journal. 2006;**48**:354-366. DOI: 10.1111/j.1365-313X.2006.02881.x

[115] Stone SL, Braybrook SA, Paula SL, Kwong LW, Meuser J, Pelletier J, et al. *Arabidopsis* LEAFY COTYLEDON2 induces maturation traits and auxin activity: Implications for somatic embryogenesis. Proceedings of the National Academy of Sciences of USA. 2008;**105**:3151-3156. DOI: 10.1073/pnas.0712364105

[116] Brady SM, McCourt P. Hormone cross-talk in seed dormancy. Journal of

Plant Growth Regulation. 2003;**22**:25-31. DOI: 10.1007/s00344-003-0018-7

[117] Braybrook SA, Harada JJ. LECs go crazy in embryo development. Trends in Plant Science. 2008;**13**:624-630. DOI: 10.1016/j.tplants.2008.09.008

[118] Gazzarrini S, Tsuchiya Y, Lumba S, Okamoto M, McCourt P. The transcription factor FUSCA3 controls developmental timing in *Arabidopsis* through the hormones gibberellin and abscisic acid. Developmental Cell. 2004;7:373-385. DOI: 10.1016/j.devcel.2004.06.017

[119] Pang Y, Wenger JP, Saathoff K, Peel GJ, Wen J, Huhman D, et al. A WD40 repeat protein from *Medicago truncatula* is necessary for tissue- specific anthocyanin and proanthocyanid in biosynthesis but not for trichome development. Plant Physiology. 2009;**151**:1114-1129. DOI: 10.1104/pp.109.144022

[120] Liu C, Jun JH, Dixon RA. MYB5 and MYB14 play pivotal roles in seed coat polymer biosynthesis in *Medicago truncatula*. Plant Physiology. 2014;**165**:1424-1439. DOI: 10.1104/pp.114.241877

[121] Verdier J, Zhao J, Torres-Jerez I, Ge S, Liu C, He X, et al. MtPAR MYB transcription factor acts as an on switch for proanthocyanidin biosynthesisin *Medicago truncatula*. Proceedings of the National Academy of Sciences of USA. 2012;**109**:1766-1771. DOI: 10.1073/pnas.1120916109

[122] Dhaubhadel S, McGarvey BD, Williams R, Gijzen M. Isoflavonoid biosynthesis and accumulation in developing soybean seeds. Plant Molecular Biology. 2003;**53**:733-743. DOI: 10.1023/B:PLAN.0000023666.30358.ae

[123] Pang Y, Peel GJ, Sharma SB, Tang Y, Dixon RA. A transcript profiling approach reveals an epicatechin-specific glucosyltransferase expressed in the seed coat of *Medicago truncatula*. Proceedings of the National Academy of Sciences of USA. 2008;**105**:14210-14215. DOI: 10.1073/pnas.0805954105

[124] Zhao J, Dixon RA. MATE transporters facilitate vacuolar uptake of epicatechin 39-O-glucoside for proanthocyanidin biosynthesis in *Medicago truncatula* and Arabidopsis. The Plant Cell. 2009;**21**:2323-2340. DOI: 10.1105/tpc.109.067819

[125] Pang Y, Cheng X, Huhman D, Ma J, Peel G, Yonekura-Sakakibara K, et al. Medicago glucosyltransferase UGT72L1: Potential roles in proanthocyanidin biosynthesis. Planta. 2013;**238**:139-154. DOI: 10.1007/s00425-013- 1879-z

[126] Zhao J, Huhman D, Shadle G, He XZ, Sumner LW, Tang YH, et al. MATE2 mediates vacuolar sequestration of flavonoid glycosides and glycoside malonates in *Medicago truncatula*. The Plant Cell. 2011;**23**:1536-1555. DOI: 10.1105/tpc.110.080804

[127] Peel GJ, Pang Y, Modolo LV, Dixon RA. The LAP1 MYB transcription factor orchestrates anthocyanidin biosynthesis and glycosylation in Medicago. The Plant Journal. 2009;**59**:136-149. DOI: 10.1111/j.1365-313X.2009.03885.x

[128] Locascio A, Roig-Villanova I, Bernardi J, Varotto S. Current perspectives on the hormonal control of seed development in Arabidopsis and maize: A focus on auxin. Frontiers in Plant Science. 2014;**5**:412. DOI: 10.3389/fpls.2014.00412

[129] Yang J, Zhang J. Grain filling of cereals under soil drying.

New Phytologist. 2006;**169**:223-236. DOI: 10.1111/j.1469-8137.2005.01597.x

[130] Ahmad I, Kamran M, Meng X, Ali S, Ahmad S, Gao Z, et al. Hormonal changes with uniconazole trigger canopy apparent photosynthesis and grain filling in wheat crop in a semi-arid climate. Protoplasma. 2021;**258**:139-150. DOI: 10.1007/s00709-020-01559-0

[131] Kong L, Guo H, Sun M. Signal transduction during wheat grain development. Planta. 2015;**241**:789-801. DOI: 10.1007/s00425-015-2260-1

[132] Lv XK, Li T, Wen XX, Liao YC, Liu Y. Effect of potassium foliage application post-anthesis on grain filling of wheat under drought stress. Field Crops Research. 2017;**206**:95-105. DOI: 10.1016/j.fcr.2017.02.015

[133] Ahmad I, Kamran M, Ali S, Cai T, Bilegjargal B, Liu T, et al. Seed filling in maize and hormones crosstalk regulated by exogenous application of uniconazole in semiarid regions. Environmental Science and Pollution Research. 2018;**25**:33225-33239. DOI: 10.1007/s11356-018-3235-0

[134] Nambara E, Marion-Poll A. Abscisic acid biosynthesis and catabolism. Annual Review of Plant Biology. 2005;**56**:165-185. DOI: 10.1146/annurev.arplant.56.032604.144046

[135] Qin SJ, Zhang ZZ, Ning TY, Ren SZ, Su LC, Li ZJ. Abscisic acid and aldehyde oxidase activity in maize ear leaf and grain relative to postflowering photosynthetic capacity and grain-filling rate under different water/nitrogen treatments. Plant Physiology and Biochemistry. 2013;**70**:69-80. DOI: 10.1016/j.plaphy.2013.04.024

[136] Liu Y, Han J, Liu DD, Gu DD, Wang YP, Liao YC, et al. Effect of plastic film mulching on the grain filling and hormonal changes of maize under different irrigation conditions. PLoS One. 2015;**10**:e0122791. DOI: 10.1371/journal.pone.0122791

[137] Chen K, Ali S, Chen Y, Manzoor SA, Jan A, Inamullah FS. Effect of ridge-covering mulching materials on hormonal changes, antioxidative enzyme activities and production of maize in semi-arid regions of China. Agricultural Water Management. 2018;**204**:281-291. DOI: 10.1016/j.agwat.2018.03.023

[138] Yu SM, Lo SF, Ho THD. Source–sink communication: Regulated by hormone, nutrient, and stress cross-signaling. Trends in Plant Science. 2015;**20**:844-857. DOI: 10.1016/j.tplants.2015.10.009

[139] Yang JC, Zhang JH, Wang ZQ, Zhu QS, Liu LJ. Involvement of abscisic acid and cytokinins in the senescence and remobilization of carbon reserves in wheat subjected to water stress during grain filling. Plant, Cell & Environment. 2003;**26**:1621-1631. DOI: 10.1046/j.1365-3040.2003.01081.x

[140] Yang J, Zhang J, Wang Z, Zhu Q, Wang W. Hormonal changes in the grains of rice subjected to water stress during grain filling. Plant Physiology. 2001;**127**:315-323. DOI: 10.1104/pp.127.1.315

[141] Zhang H, Li H, Yuan L, Wang Z, Yang J, Zhang J. Post-anthesis alternate wetting and moderate soil drying enhances activities of key enzymes in sucrose-to-starch conversion in inferior spikelets of rice. Journal of Experimental Botany. 2012;**63**:215-227. DOI: 10.1093/jxb/err263

[142] Barrôco RM, Peres A, Droual AM, De Veylder L, Nguyen LSL, De Wolf J, et al. The cyclin-dependent kinase inhibitor Orysa; KRP1 plays an

important role in seed development of rice. Plant Physiology. 2006;**142**:1053-1064. DOI: 10.1104/pp.106.087056

[143] Sreenivasulu N, Borisjuk L, Junker BH, Mock HP, Rolletschek H, Seiffert U, et al. Barley grain development: Toward an integrative view. International Review of Cell and Molecular Biology. 2010;**281**:49-89. DOI: 10.1016/S1937-6448(10)81002-0

[144] Popko J, Hänsch R, Mendel R, Polle A, Teichmann T. The role of abscisicacid and auxin in the response of poplar to abiotic stress. Plant Biology. 2010;**12**:242-258. DOI: 10.1111/j.1438-8677.2009.00305.x

[145] Liu B, Liu XB, Wang C, Jin J, Herbert SJ. Endogenous hormones in seed, leaf, and pod wall and their relationship to seed filling in soybeans. Crop and Pasture Science. 2010;**61**:103-110. DOI: 10.1071/CP09189

[146] Liu PP, Koizuka N, Homrichhausen TM, Hewitt JR, Martin RC, Nonogaki H. Large-scale screening of Arabidopsis enhancer-trap lines for seed germination-associated genes. Plant Journal. 2005;**41**: 936-944

[147] Golldack D, Li C, Mohan H, Probst N. Gibberellins and abscisic acid signal crosstalk: Living and developing under unfavorable conditions. Plant Cell Reports. 2013;**32**:1007-1016. DOI: 10.1007/s00299-013-1409-2

[148] Luján-Soto E, Dinkova TD. Time to wake up: Epigenetic and small-RNA-mediated regulation during seed germination. Plants. 2021;**10**:236. DOI: 10.3390/plants10020236

[149] Argueso CT, Ferreira FJ, Kieber JJ. Environmental perception avenues: The interaction of cytokinin and environmental response pathways. Plant, Cell & Environment. 2009;**32**:1147-1160. DOI: 10.1111/j.1365-3040.2009.01940.x

[150] Zhang W, Cao Z, Zhou Q, Chen J, Xu G, Gu J, et al. Grain filling characteristics and their relations with endogenous hormones in large- and small-grain mutants of rice. PLoS One. 2016;**11**:e0165321. DOI: 10.1371/journal.pone.0165321

[151] Morris RO, Blevins DG, Dietrich JT, Durley RC, Gelvin SB, Gray J, et al. Cytokinins in plant pathogenic bacteria and developing cereal grains. Functional Plant Biology. 1993;**20**:621-637. DOI: 10.1071/PP9930621

[152] Yang J, Peng S, Visperas RM, Sanico AL, Zhu Q, Gu S. Grain filling pattern and cytokinin content in the grains and roots of rice plants. Plant Growth Regulation. 2000;**30**:261-270. DOI: 10.1023/A:1006356125418

[153] Ahmad I, Kamran M, Meng X, Ali S, Bilegjargal B, Cai T, et al. Effects of plant growth regulators on seed filling, endogenous hormone contents and maize production in semiarid regions. Journal of Plant Growth Regulation. 2019;**38**:1467-1480. DOI: 10.1007/s00344-019-09949-2

[154] Morris RO. Hormonal regulation of seed development. In: Larkins BA, Vasil IK, editors. Cellular and Molecular Biology of Plant Seed Development. Dordrecht: Springer; 1997. pp. 117-149. DOI: 10.1007/978-94-015-8909-3_4

[155] Cui W, Song Q, Zuo B, Han Q, Jia Z. Effects of gibberellin (GA_{4+7}) in grain filling, hormonal behavior, and antioxidants in high-density maize (*Zea mays* L.). Plants. 2020;**9**:978. DOI: 10.3390/plants9080978

[156] Jasinski S, Piazza P, Craft J, Hay A, Woolley L, Rieu I, et al. KNOX action in

Arabidopsis is mediated by coordinate regulation of cytokinin and gibberellin activities. Current Biology. 2005;**15**:1560-1565. DOI: 10.1016/j.cub.2005.07.023

[157] Xu YJ, Gu DJ, Zhang BB, Zhang H, Wang ZQ, Yang JC. Hormone contents in kernels at different positions on an ear and their relationship with endosperm development and kernel filling in maize. Acta Agronomica Sinica. 2013;**39**:1452-1461. DOI: 10.3724/SP.J.1006.2013.01452

[158] Liu Y, Han J, Wen XX, Wu W, Guo Q, Zeng AI, et al. Effect of plastic-covered ridge and furrow planting on grain filling of winter wheat is related to changes in endogenous hormones. Journal of Integrative Agriculture. 2013;**12**:1771-1782. DOI: 10.1016/S2095-3119(13)60337-8

[159] Liu Y, Sui YW, Gu DD, Wen XX, Chen Y, Li CJ, et al. Effects of conservation tillage on grain filling and hormonal changes in wheat under simulated rainfall conditions. Field Crops Research. 2013;**144**:43-51. DOI: 10.1016/j.fcr.2013.01.009

[160] Liu Y, Peng L, Dong ST, Zhang JW. Hormonal changes caused by the xenia effect during grain filling of normal corn and high-oil corn crosses. Crop Science. 2010;**50**:215-221. DOI: 10.2135/cropsci2009.04.0186

[161] Jia Q, Sun L, Mou H, Ali S, Liu D, Zhang Y, et al. Effects of planting patterns and sowing densities on grain-filling, radiation use efficiency and yield of maize (*Zea mays* L.) in semiarid regions. Agricultural Water Management. 2018;**201**:287-298. DOI: 10.1016/j.agwat.2017.11.025

[162] Tamas IA, Engels CJ, Kaplan SL, Ozbun JL, Wallace DH. Role of indoleacetic acid and abscisic acid in the correlative control by fruits of axillary bud development and leaf senescence. Plant Physiology. 1981;**68**:476-481. DOI: 10.1104/pp.68.2.476

[163] Singh G, Gerung SB. Hormonal role in the problem of sterility in *Oryza sativa*. Plant Physiology Biochemistry. 1982;**9**:22-23

[164] Ahmad S, Wennan S, Kamran M, Ahmad I, Meng X, Xiaorong W, et al. Foliar application of melatonin delay leaf senescence in maize by improving the antioxidant defense system and enhancing photosynthetic capacity under semiarid regions. Protoplasma. 2020;**257**:1079-1092

[165] Rijavec T, Kovaˇc M, Kladnik A, Chourey PS, Dermastia M. Comparative study on the role of cytokinins in caryopsis development in the maize miniature1 seed mutant and its wild type. Journal of Integrative Plant Biology. 2009;**51**:840-849. DOI: 10.1111/j.1744-7909.2009.00863.x

[166] Panda BB, Kariali E, Panigrahi R, Mohapatra PK. High ethylene production slackens seed filling in compact panicled rice cultivar. Plant Growth Regulation. 2009;**58**:141-151. DOI: 10.1007/s10725-009-9362-3

[167] Kudryakova NV, Burkhanova EA, Rakitin VY, Yakovleva LA, Smith A, Hall MA, et al. Ethylene and cytokinin in the control of senescence in detached leaves of Arabidopsis thaliana eti-5mutant and wild-type plants. Russian Journal of Plant Physiology. 2001;**48**:624-627. DOI: 10.1023/A:1016708002899

[168] Taverner E, Letham DS, Wang J, Cornish E, Willcocks DA. Influence of ethylene on cytokinin metabolism in relation to Petunia corolla senescence. Phytochemistry. 1999;**51**:341-347

[169] Ochatt SJ. Agroecological impact of an in vitro biotechnology approach

of embryo development and seed filling in legumes. Agronomy for Sustainable Development. 2015;**35**:535-552. DOI: 10.1007/s13593-014-0276-8

[170] Lichtenhaler HK. Vegatation stress: An introduction to the stress concept in plants. Journal of Plant Physiology. 1996;**148**:4-14

[171] Zinn KE, Tunc-Ozdemir M, Harper JF. Temperature stres and plan sexual reproduction: Uncovering the weakest links. Journal Experimental Botany. 2010;**61**:1959-1968

[172] Thakur P, Kumar S, Malik JA, Berger JD, Nayyar H. Cold stress effects on reproductive development in grain crops: An overview. Environmental and Experimental Botany. 2010;**67**:429-443. DOI: 10.1016/j.envexpbot.209.09.004

[173] Kumar S, Malik J, Thakur P, Kaistha S, Sharma KD, Upadhyaya HD, et al. Growth and metabolic responses of contrasting chickpea (*Cicer arietinum* L.) genotypes to chiling stress at reproductive phase. Acta Physiologiae Plantarum. 2011;**33**:779-787. DOI: 10.1007/s11738-010-0602-y

[174] Prasad PVV, Pisipati SR, Mutava RN, Tuinstra MR. Sensitivity of grain sorghum to high temperature stress during reproductive development. Crop Science. 2008;**48**:1911-1917. DOI: 10.1016/j.jplph.2009.11.007

[175] Sehgal A, Sita K, Kumar J, Kumar S, Singh S, Siddique KHM, et al. Effects of drought, heat and their interaction on the growth, yield and photosynthetic function of lentil (Lens culinaris Medikus) genotypes varying in heat and drought sensitivity. Frontiers in Plant Science. 2017;**8**:1776. DOI: 10.3389/fpls.2017. 01776

[176] Asseng S, van Herwaarden AF. Analysis of the benefits to wheat yield from assimilates stored prior to grain filling in a range of environments. Plant and Soil. 2003;**256**:217-229. DOI: 10.1023/A:1026231904221

[177] Plaut Z, Butow BJ, Blumenthal CS, Wrigley CW. Transport of dry matter into developing wheat kernels and its contribution to grain yield under post-anthesis water deficit and elevated temperature. Field Crops Research. 2004;**86**:185-198. DOI: 10.1016/j.fcr.2003.08.005

[178] Anjum SA, Xie XY, Wang LC, Saleem MF, Man C, Lei W. Morphological, physiological and biochemical responses of plants to drought stress. African Journal of Agricultural Research. 2011;**6**:2026-2032

[179] Siddique KHM, Regan KL, Tennant G, Thomson BD. Water use and water use efficiency of cool season grain legumes in low rainfall Mediterranean-type environments. European Journal of Agronomy. 2001;**15**:267-280. DOI: 10.1016/ S1161-0301(01)00106-X

[180] Valentovic P, Luxova M, Kolarovic L, Gasparikova O. Effect of osmotic stress on compatible solutes content, membrane stability and water relations in two maize cultivars. Plant, Soil and Environment. 2006;**52**:184-191

[181] França MGC, Thi ATP, Pimentel C, Rossiello ROP, Zuily-Fodil Y, Laffray D. Differences in growth and water relations among Phaseolus vulgaris cultivars in response to induced drought stress. Environmental and Experimental Botany. 2000;**43**:227-237. DOI: 10.1016/ S0098-8472(99)00060-X

[182] Farooq M, Wahid A, Kobayashi N, Fujita D, Basra SMA. Plant drought stress: Effects, mechanisms and management. Agronomy for Sustainable

Development. 2009;**29**:185-212. DOI: 10.1051/agro:2008021

[183] Sita K, Sehgal A, Kumar J, Kumar S, Singh S, Siddique KHM, et al. Identification of high-temperature tolerant lentil (*Lens culinaris* Medik.) genotypes through leaf and pollen traits. Frontiers in. Plant Science. 2017;**8**:744. DOI: 10.3389/ fpls.2017.00744

[184] Massacci A, Nabiev SM, Pietrosanti L, Nematov SK, Chernikova TN, Thor K, et al. Response of the photosynthetic apparatus of cotton (*Gossypium hirsutum*) to the onset of drought stress under field conditions studied by gas-exchange analysis and chlorophyll fluorescence imaging. Plant Physiology and Biochemistry. 2008;**46**:189-195. DOI: 10.1016/j. plaphy.2007. 10.006

[185] Rahbarian R, Khavari-Nejad R, Ganjeali A, Bagheri A, Najafi F. Drought stress effects on photosynthesis, chlorophyll fluorescence and water relations in tolerant and susceptible chickpea genotypes. Acta Biologica Cracoviensia Series Botanica. 2011;**53**:47-56. DOI: 10.2478/v10182-011-0007-2

[186] Awasthi R, Kaushal N, Vadez V, Turner NC, Berger J, Siddique KH, et al. Individual and combined effects of transient drought and heat stress on carbon assimilation and seed filling in chickpea. Functional Plant Biology. 2014;**41**:1148-1167. DOI: 10.1071/FP13340

[187] Samarah NH, Alqudah AM, Amayreh JA, McAndrews GM. The effect of late-terminal drought stress on yield components of four barley cultivars. Journal of Agronomy and Crop Science. 2009;**195**:427-441. DOI: 10.1111/j.1439-037X.2009. 00387.x

[188] Jiang Y, Huang B. Drought and heat stress injury to two cool season turfgrasses in relation to antioxidant metabolism and lipid peroxidation. Crop Science. 2001;**41**:436-442. DOI: 10.2135/ cropsci2001.412436x

[189] Samarah N, Mullen R, Cianzio S. Size distribution and mineral nutrients of soybean seeds in response to drought stress. Journal of Plant Nutrition. 2004;**27**:815-835. DOI: 10.1081/ PLN-120030673

[190] Gunes A, Cicel N, Inal A, Alpaslan M, Eraslan F, Guneri E, et al. Genotypic response of chickpea (*Cicer arietinum* L.) cultivars to drought stress implemented at pre- and post-anthesis stages and its relations with nutrient uptake and efficiency. Plant Soil Environment. 2006;**52**:368-376. DOI: 10.17221/3454-PSE

[191] Serraj R. Effects of drought stress on legume symbiotic nitrogen fixation: Physiological mechanisms. Indian Journal of Experimental Biology. 2003;**41**:1136-1141

[192] Gonzalez A, Galvez L, Royuela M, Aparicio-Tejo P, Arrese-Igor C. Insights into the regulation of nitrogen fixation in pea nodules: Lessons from drought, abscisic acid and increased assimilate availability. Agronomie. 2001;**21**:607-613. DOI: 10.1051/agro:2001151

[193] Leport L, Turner NC, Davies SL, Siddique KHM. Variation in pod production and abortion among chickpea cultivars under terminal drought. European Journal of Agronomy. 2006;**24**:236-246. DOI: 10.1016/j. eja.2005.08.005

[194] Mafakheri A, Siosemardeh A, Bahramnejad B, Struik PC, Sohrabi Y. Effect of drought stress on yield, proline and chlorophyll contents in three chickpea cultivars. Australian Journal of Crop Science. 2010;**4**:580-585

[195] Zare M, Ghahremaninejad M, Bazrafshan F. Influence of drought stress on some traits in five mung bean (*Vigna radiata* (L.) R. Wilczek) genotypes. International Journal of Agronomy and Plant Production. 2012;**3**:234-240

[196] Seghatoleslami MJ, Kafi M, Majidi E. Effect of drought stress at different growth stages on yield and water use efficiency of five proso millet (*Panicum miliaceum* L.) genotypes. Pakistan Journal of Botany. 2008;**40**:1427-1432

[197] Pushpavalli R, Zaman-Allah M, Turner NC, Baddam R, Rao MV, Vadez V. Higher flower and seed number leads to higher yield under water stress conditions imposed during reproduction in chickpea. Functional Plant Biology. 2014;**42**:162-174. DOI: 10.1071/FP14135

[198] Farooq M, Hussain M, Wakeel A, Siddique KHM. Salt stress in maize: Effects, resistance mechanisms and management. A review. Agronomy for Sustainable Development. 2015;**35**:461-481. DOI: 10.1007/s13593-015-0287-0

[199] Liu F, Jensen CR, Andersen MN. Pod set related to photosynthetic rate and endogenous ABA in soybeans subjected to different water regimes and exogenous ABA and BA at early reproductive stages. Annals of Botany. 2004;**94**:405-411. DOI: 10.1093/aob/mch157

[200] Saeedipour S. Activities of sucrose-metabolizing enzymes in grains of two wheat (*Triticum aestivum* L.) cultivars subjected to water stress during grain filling. Journal of Plant Breeding and Crop Science. 2011;**3**:106-113

[201] Shrestha R, Turner NC, Siddique KHM, Turner DW, Speijers J. A water deficit during pod development in lentils reduces flower and pod numbers but not seed size. Australian Journal of Agricultural Research. 2006;**57**:427-438. DOI: 10.1071/ AR05225

[202] Ghanbari AA, Mousavi SH, Mousapour Gorgi A, Rao IM. Effects of water stress on leaves and seeds of bean (*Phaseolus vulgaris* L.). Turkish Journal of Field Crops. 2013;**181**:73-77

[203] Smiciklas KD, Mullen RE, Carlson RE, Knapp AD. Soybean seed quality response to drought stress and pod position. Agronomy Journal. 1992;**84**:166-170. DOI: 10.2134/agronj1992.00021962008400020008x

[204] Samarah NH, Haddad N, Alqudah A. Yield potential evaluation in chickpea genotypes under late terminal drought in relation to the length of reproductive stage. Italian Journal of Agronomy. 2009;**3**:111-117. DOI: 10.4081/ija.2009. 3.111

[205] Saini HS, Westgate ME. Reproductive development in grain crops during drought. In: Sparks DL, editor. Advances in Agronomy. Vol 68. Delaware: Academic Press; 1999. p. 59-96. DOI: 10.1016/S0065-2113(08)60843-3

[206] Nicolas ME, Gleadow RM, Dalling MJ. Effect of postanthesis drought on cell division and starch accumulation in developing wheat grains. Annals of Botany. 1985;**55**:433-444. DOI: 10.1093/oxfordjournals.aob.a08 6922

[207] Behboudian MH, Ma Q, Turner NC, Palta JA. Reactions of chickpea to water stress: Yield and seed composition. Journal of the Science of Food and Agriculture. 2001;**81**:1288-1291. DOI: 10.1002/jsfa.939

[208] Ahmadi A, Baker DA. The effect of water stress on the sucrose to starch pathway. Plant Growth Regulation. 2001;**35**:81-91. DOI: 10.1023/A:1013827600528

[209] Singh S, Singh G, Singh P, Singh N. Effect of water stress at different stages of grain development on the characteristics of starch and protein of different wheat varieties. Food Chemistry. 2008;**108**:130-139. DOI: 10.1016/j.foodchem. 2007.10.054

[210] Bellaloui N, Hu Y, Mengistu A, Kassem MA, Abel CA. Effects of foliar boron application on seed composition, cell wall boron, and seed _15N and _13C isotopes in water-stressed soybean plants. Frontiers in Plant Science. 2013;**4**:270. DOI: 10.3389/fpls.2013.00270

[211] Dwivedi SL, Nigam SN, Rao RCN, Singh U, Rao KVS. Effect of drought on oil, fatty acids and protein contents of groundnut (*Arachis hypogaea* L.) seeds. Field Crops Research. 1996;**48**:125-133. DOI: 10.1016/S0378-4290(96) 01027-1

[212] Ali Q, Ashraf M, Anwar F. Seed composition and seed oil antioxidant activity of maize under water stress. Journal of the American Oil Chemists Society. 2010;**87**:1179-1187. DOI: 10.1007/s11746-010-1599-5

[213] Ali Q, Ashraf M, Anwar F, Al-Qurainy F. Trehalose-induced changes in seed oil composition and antioxidant potential of maize grown under drought stress. Journal of the American Oil Chemists Society. 2012;**89**:1485-1493. DOI: 10.1007/s11746- 012-2032-z

[214] Dornbos DLJr, Mullen RE. Soybean seed protein and oil contents and fatty-acid composition adjustments by drought and temperature. Journal of American Oil Chemists' Society. 1992;**69**:228-231. DOI: 10.1007/BF02635891

[215] Sung DY, Kaplan F, Lee KJ, Guy CL. Acquired tolerance to temperature extremes. Trends Plant Science. 2003;**8**:179-187. DOI: 10.1016/S1360-1385(03)00047-5

[216] Kaushal N, Bhandari K, Siddique KHM, Nayyar H. Food crops face rising temperatures: An overview of responses, adaptive mechanisms, and approaches to improve heat tolerance. Cogent Food Agriculture. 2016;**2**:1134380. DOI: 10.1080/23311932. 2015.1134380

[217] Dias AS, Lidon FC. Evaluation of grain filling rate and duration in bread and durum wheat, under heat stress after anthesis. Journal of Agronom Crop Science. 2009;**195**:137-147. DOI: 10.1111/j.1439-037X.2008.00347.x

[218] Prasad PVV, Pisipati SR, Momcilovic I, Ristic Z. Independent and combined effects of high temperature and drought stress during grain filling on plant yield and chloroplast protein synthesis elongation factor (EFTu) expression in spring wheat. Journal of Agronomy and Crop Science. 2011;**197**:430, 10.1111/j. 1439-037X.2011.00477.x-441

[219] Prasad PVV, Craufurd PQ, Summerfield RJ, Wheeler TR. Effects of short episodes of heat stress on flower production and fruit-set of groundnut (*Arachis hypogaea* L.). Journal of Experimental Botany. 2000;**51**:777-784. DOI: 10.1093/jexbot/ 51.345.777

[220] Tsukaguchi T, Kawamitsu Y, Takeda H, Suzuki K, Egawa Y. Water status of flower buds and leaves as affected by high temperature in contrasting snap bean (*Phaseolus vulgaris* L.) varieties in heat tolerance. Plant Production Science. 2003;**6**:24-27. DOI: 10.1626/pps.6.24

[221] Devasirvatham V, Tan DKY, Trethowan RM, Gaur PM, Mallikarjuna N. Impact of high temperature on the reproductive stage of chickpea. In: Food Security from Sustainable Agriculture Proceedings of

the 15th Australian Society of Agronomy Conference. New Zealand: Lincoln; 2010. pp. 15-18

[222] Djanaguiraman M, Prasad PV, Seppanen M. Selenium protects sorghum leaves from oxidative damage under high temperature stress by enhancing antioxidant defense system. Plant Physiology and Biochemistry. 2010;**48**:999-1007

[223] Rashid M, Hampton JG, Rolston MP, Khan KM, Saville DJ. Heat stress during seed development affects forage brassica (*Brassica napus* L.) seed quality. Journal of Agronomy Crop Science. 2017;**204**:147-154. DOI: 10.1111/jac.12251

[224] Kaushal N, Awasthi R, Gupta K, Gaur P, Siddique KH, Nayyar H. Heat-stress-induced reproductive failures in chickpea (*Cicer arietinum*) are associated with impaired sucrose metabolism in leaves and anthers. Functional Plant Biology. 2013;**40**:1334-1349. DOI: 10.1071/FP13082

[225] Duthion C, Pigeaire A. Seed lengths corresponding to the final stage in seed abortion of three grain legumes. Crop Science. 1991;**31**:1579-1583. DOI: 10.2135/cropsci1991.0011183X003100060040x

[226] Nielsen CL, Hall AE. Responses of cowpea (*Vigna unguiculata* (L.)Walp.) in the field to high night air temperature during flowering. II. Plant responses. Field Crops Research. 1985;**10**:181-196. DOI: 10.1016/0378-4290(85)90025-5

[227] Chakrabarti B, Singh SD, Kumar V, Harit RC, Misra S. Growth and yield response of wheat and chickpea crops under high temperature. Indian Journal of Plant Physiology. 2013;**18**:7-14. DOI: 10.1007/s40502-013-0002-6

[228] Barnabás B, Jäger K, Fehér A. The effect of drought and heat stress on reproductive processes in cereals. Plant Cell Environment. 2008;**31**:11-38

[229] Hurkman WJ, McCue KF, Altenbach SB, Korn A, Tanaka CK, Kothari KM, et al. Effect of temperature on expression of genes encoding enzymes for starch biosynthesis in developing wheat endosperm. Plant Science. 2003;**164**:873-881. DOI: 10.1016/S0168-9452(03)00076-1

[230] Yamakawa H, Hakata M. Atlas of rice grain filling-related metabolism under high temperature: Joint analysis of metabolome and transcriptome demonstrated inhibition of starch accumulation and induction of amino-acid accumulation. Plant Cell Physiology. 2010;**51**:795-809. DOI: 10.1093/pcp/pcq034

[231] DuPont FM, Altenbach SB. Molecular and biochemical impacts of environmental factors on wheat grain development and protein synthesis. Journal of Cereal Science. 2003;**38**:133-146. DOI: 10.1016/S0733-5210(03)00030-4

[232] Thomas JMG, Boote KJ, Allen LH, Gallo-Meagher M, Davis JM. Elevated temperature and carbon dioxide effects on soybean seed germination and transcript abundance. Crop Science. 2003;**43**:1548-1557. DOI: 10.2135/cropsci2003.1548

[233] Jenner CF. Effects of exposure of wheat ears to high temperature on dry matter accumulation and carbohydrate metabolism in the grain of two cultivars. Functional Plant Biology. 1991;**18**:165-177. DOI: 10.1071/PP9910165

[234] Yamakawa H, Hirose T, Kuroda M, Yamaguchi T. Comprehensive expression profiling of rice grain filling-related genes under high temperature using DNA microarray. Plant Physiology.

2007;**144**:258-277. DOI: 10.1104/pp.107.098665

[235] Tanamachi K, Miyazaki M, Matsuo K, Suriyasak C, Tamada A, Matsuyama K, et al. Differential responses to high temperature during maturation in heat-stress tolerant cultivars of japonica rice. Plant Production Science. 2016;**19**:300-308. DOI: 10.1080/1343943X.2016.1140007

[236] Ozkaynak E, Samancı B. Comparison of popcorn (*Zea mays* everta Sturt.) lines and their testcrosses for yield and yield-related traits. Mediterranean Agricultural Sciences. 2003;**16**(1):35-42

[237] Bartholomew SB. Temperature effects on the fatty acid composition of developing seeds in safflower, *Carthamus tinctorius* S.L. [thesis]. University of California, Davis; 1971

[238] Canvin DT. The effect of temperature on the oil content and fatty acid composition of the oils from several oil seed crops. Canadian Journal of Botany. 1965;**43**:63-69. DOI: 10.1139/b65-008

[239] Brunel-Muguet S, D'Hooghe P, Bataillé MP, Larré C, Kim TH, Trouverie J, et al. Heat stress during seed filling interferes with sulfur restriction on grain composition and seed germination in oilseed rape (*Brassica napus* L.). Frontiers in Plant Science. 2015;**6**:213. DOI: 10.3389/fpls.2015.00213

[240] Weiss EA. Oilseed Crops. Leonord Hill Boks, New York: Tropical Agriculture Series. Pub. By Longman Inc.; 1983

[241] Röbbelen G, Downey RK, Ashri A. Oilcrops of the World. USA: McgrawHill; 1989

[242] Erbaş S, Tonguç M, Şanlı A. Variations in the agronomic and quality characteristics of domestic and foreign safflower *Carthamus tinctorius* L genotypes. Turkish Journal of Field Crops. 2016;**21**:110-119. DOI: 10.17557/tjfc.56268

[243] Pritchard FM, Eagles HA, Norton RM, Salisbury PA, Nicolas M. Environmental effects on seed composition of *Victorian canola*. Australian Journal of Experimental Agriculture. 2006;**40**(5):679-685

[244] Broun P, Somerville C. Accumulation of Ricinoleic, Lesquerolic and Densipolic acid an seeds to transgenic Arabidopsis plants that express a fatty acylhdroxlase from Castor bean. Plant Phsiology. 1997;**113**:933-942

[245] Stryer L. Biochemistry. 30th press. New York: W. H. Freeman Comp. Inc.; 1986

[246] Dybing CD, Zimmerman DC. Fatty acid accumulation in maturing flax seeds as influenced by environment. Plant Physiology. 1966;**41**:1465-1470

[247] Harris HC, McWilliam JR, Mason WK. Influence of temperature on oil content and composition of sunflower seed. Australian Journal of Agricultural Research. 1978;**29**:1203-1212. DOI: 10.1071/AR9781203

[248] Stone PJ, Gras PW, Nicolas ME. The influence of recovery temperature on the effects of brief heat shock on wheat. III. Grain protein and dough properties. Journal of Cereal Science. 1997;**25**:129-141. DOI: 10.1006/jcrs.1996.0080

[249] Triboï E, Martre P, Triboï-Blondel AM. Environmentally-induced changes of protein composition for developing grains of wheat are related to changes in total protein

content. Journal of Experimental Botany. 2003;**54**:1731-1742. DOI: 10.1093/jxb/erg183

[250] Mession JL, Sok N, Assifaoui A, Saurel R. Thermal denaturation of pea globulins (*pisum sativum* L.) molecular interactions leading to heatinduced protein aggregation. Journal of Agriculture and Food Chemistry. 2013;**61**:1196-1204. DOI: 10.1021/jf303739n

[251] Zu L. Effects of gradual and sudden heat stress on seed quality of Andean Lupin, *Lupinus mutabilis* [thesis]. Helsinki: The University of Helsinki; 2009

Chapter 2

Seed Biology and Phytochemistry for Sustainable Future

Nikhat Farhana, Ripudaman M. Singh, Mohammed Gulzar Ahmed, Thouheed Ansari, Abdul Rahamanulla, Ayesha Sultana, Treesa P. Varghese, Ashwini Somayaji and Abdullah Khan

Abstract

The ranking of seeds represents remarkable transition phase for photosynthasis and sexual reproduction, this phase is complex & successful method for sexual reproduction in vascular plants. As we know that seed containe the genetic repository of past & potential for its perpetuation in the future. The dormancy in seeds induced by desciccation & the hormone abscisic acid (ABA) till the condition in growth become favorable. The well developed seeds eliminates requirement of water during sexual reproduction & allows fertilization events to occur over long distances. Germination of seeds in particular situation and season is determines the interaction between dormancy and relating factors like phytochemical development to give healthy bioactives, which strongly influence on the termination of dormancy or initiatin of germination and seedling in many plant species like photo-hormons, light, temperature, water, neutrients and mechanical cuse. Seeds of particular plants needs difreent pretreatment to give vigor seedlings even in production so far. The entiteled chapter represents amulgumation of agriculturists and life scientists. Recent significant progress has been endorsed in seed physiology to solve the practical issues constantly associated with the seeds. The aim & objective of this articale is to enlighten the reader, not only about the different aspects of the seed physiology it also includes the development of bioactive (secondary metabolites) in the healthy seeds. This resorce of pice will help researcher to sencitize about the type of healthy bioactive available in the shalls of seedlings. This could be the reason to isolate the biomolecules from a well evaluate seeds, seed evalution not only the sorce to get healthy crops in agricultural scince it also helps so for a phytochemist to get theuraputicaly active biomolecules, without destroying the nature, which could be the value added thought to combat with the burning issues associated with the existing situation (COVID Omicron, viral infection and all kinds of disorder associated with the immune system). Henceforth, endorsed personage to give real-time attention to plant propagation, particularly for indigenous tree species and seedling multiplication should be regarded as a primary need to make not only a sustainable environment but also become a treasure to fulfill the needs of industry application in the field of agriculture plus R & D.

IntechOpen

Keywords: germination, synthesis, natural molecules, phytochemistry, industry, endorsed, COVID Omicron, R & D

1. Introduction

The Life cycle of flowering plants follows distinct development milestones which start beginning from seeds to converting into plants or crops, ultimately enabling new seed production, plants are adaptive to the native environment, and in the time taken from fertilization and germination, anything can be done to increase the proportion of seeds that emerge and the rate at which they do so, will have a large impact on farmers and researcher's livelihood, a sensitized former knows about mechanisms underlaying, development germinability, dormancy and storability to improve the performance of seeds which involves temperature, moisture, oxygen light and all other factors related to storage similarly the natural chemistry research depends on the perceptions of formers working attribute to make healthy seeds which come out with highly active primary and secondary metabolites. Seeds are considered a major source of food hence all information concerning their nutritive value, chemical composition and quality. Several hormones and chemicals are used to improve the oil, protein, and other economic attributes of seeds. Overall to say seeds are the connection between the past and future. They contain the genetic wisdom of the past and the potential for its perpetuation in future. The natural packaging of the genetic repository remarkably protects the germplasm collection. This chapter takes the reader to the world of healthy seeds with its repository of chemical composition required and enlighten the reader about the biotechnological research during the last two decade and opened up unprecedented opportunities in any area of basic and applied biological research, plant tissue culture which is important components of plant biotechnology, phytochemistry and pharmacological importance put ups the new strategy for the improvements of cereals, legumes, forest trees, crops plantation, ornamental plant. Nowadays, seed technology is a most important tool to breeders and scientists of plant tissue culture and phytochemistry, it has offered a powerful advantage for large-scale mass propagation of elite species [1].

2. Germination in the soil and standard establishment

Germination is the fundamental process by which different plant species grow from a single seed into a young plant. This process essentially influences both crops yield and quality, here the seed observes water by the passage of time, chilling, warming, oxygen availability, and light exposure may all operate while initiating the process, the rehydration will expand the cell embryo which increases the rate of respiration, and various metabolic processes suspended or much reduced during dormancy resume. These events in the life of seeds are associated with anatomical changes in cell organelles (membranous bodies concerned with metabolism), in the cell of the embryo inside the stigma, whereas each seed reacts individually to its microenvironment. A field consists of a wide range of microenvironments, how seeds establish & germinate under field conditions hence the uniformity of crops associated with seedbed preparation (perry 1973; Hadas wolf), the performance of crops' physical and chemical properties is as per the stability climatic uncertainty and traffic history. Hence germination is the preliminary stage to represent the quality of seeds (**Figure 1**) [2].

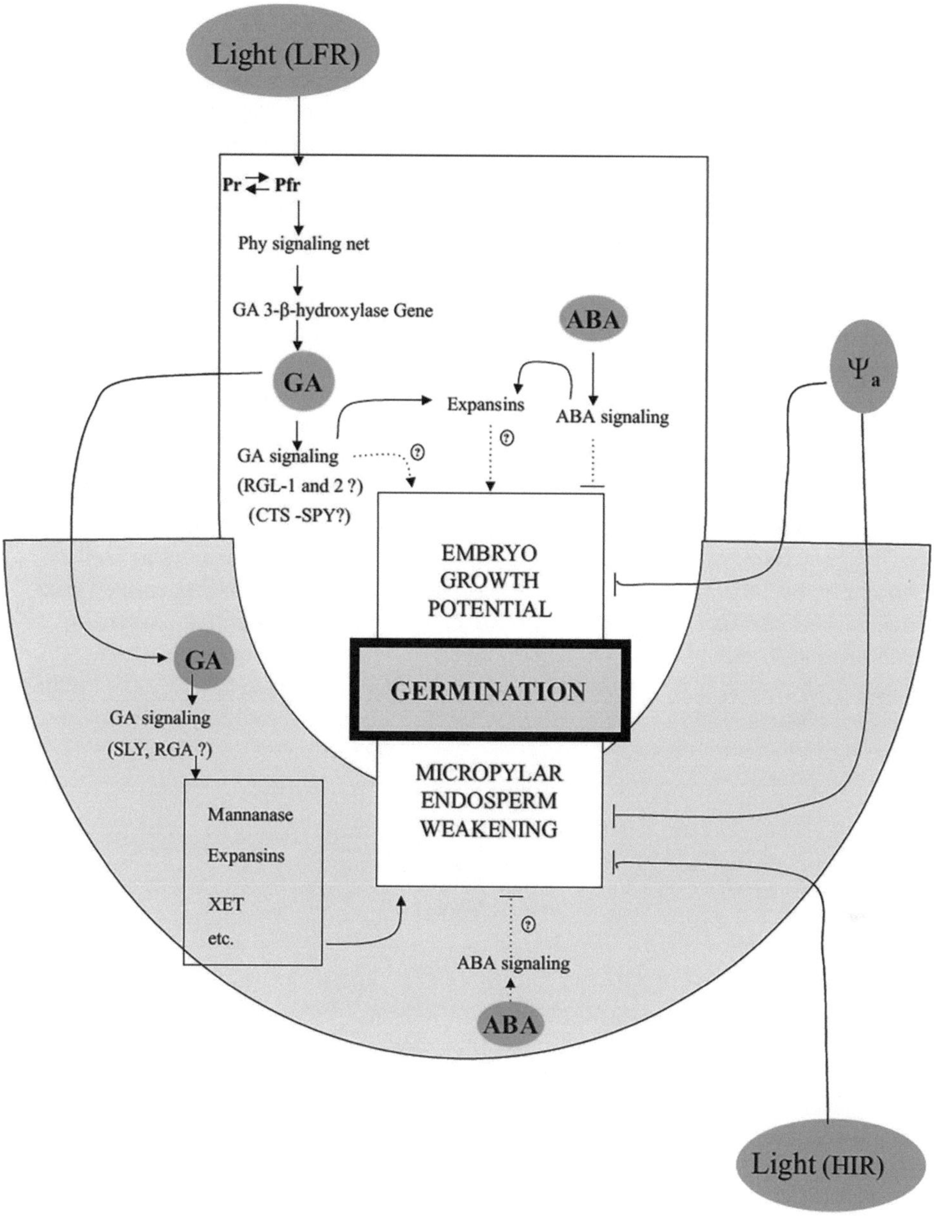

Figure 1.
Diagrammatic representation of part of the molecular signaling and components acting on the termination of coat-imposed dormancy and the induction of germination. Broken lines and question marks represent probable but unconfirmed interactions.

3. Dormancy of the crops and weeds

As discussed, the dormancy of crops & weeds goes on the season. Dormancy is the state of seeds and buds were alive but not germinating if the process takes place once the seedlings get destroyed. Dormancy allows the storage of millions of seeds in the soil and enables them to grow in flushes over the years. In this context the horticulturist saying will fit onto "One-year seeding seven-year weeding" appropriately fit onto the heading, the high persistence of seeds results from their multifaced mechanism important among these are; prolific seed production, vegetative propagation, rapid dispersal, inherent hardiness, evasiveness, self-regeneration, selective invention and weed success. Whereas weeds and seeds are dormant for three reasons i.e. enforced dormancy, innate dormancy and induced dormancy. Finally, the overall persistence of weed depends upon its capability to adopt one or more of the above-cited features. A weed species that embodied the majority of these factors is surely a horrible weed (**Figure 2**).

To conclude seed dormancy has different elaborations based on the different lookout and thinking of beings, this was a highly complicated phenomenon and weakly understood even though a huge number of publications available on this topic, as mentioned factors above the complexity are due to mechanical, physiological and biological some time it may be controlled by the environment. The known fact of dormancy not only induction and braking but complications were interrelated issues with the seed's anatomy and physiology like seed coat, embryo, cotyledon, endosperm, cell organelles, nuclei, all need much research with the role of external environment on seeds. Weeds are of most concern to farmers as well as researchers, backup data is available on this but less research relating to seed coat structure, temperature, pressure, light, hormones synthetic chemical enzymes metabolites and related chemical factors need to be explored, however seed dormancy and weed is the main issues needs in-depth research to solve the formers researchers issues related to field environment or chemical composition both are interrelated vice and versa [3].

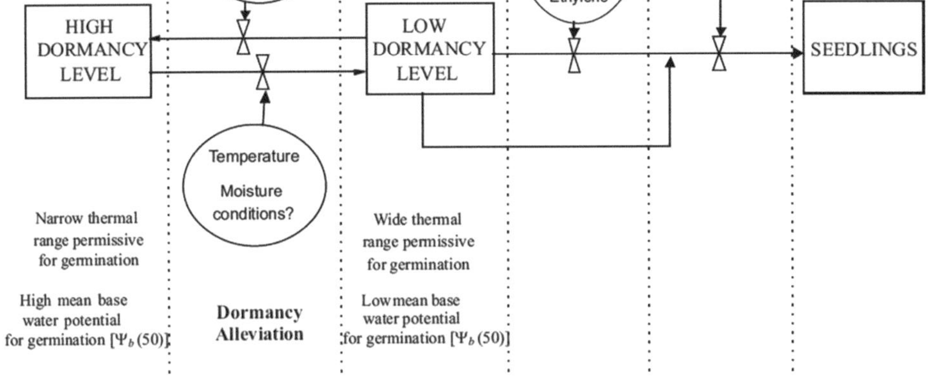

Figure 2.
Flowchart representing changes in dormancy level and termination of dormancy in seed populations and the factors that most likely affect each process (source: Reprinted from field crops research 67 [2], R. L. Benech-Arnoldet al., environmental control of dormancy in weed seed banks in soil, pp. 105–122, copyright 2000, with permission from Elsevier science).

4. Seeds longevity and storage

Seeds even if adequately protected while storage the chances are more to undergo deterioration with time, major factors affecting the longevity (life span) of mature, variable and healthy seeds are moisture, storage, temperature and pests. Seeds are of different species require different storage conditions, some seeds with the hard court can be stored at room temperature for several years (42°F to 5.6°C), the actual storage is depending upon the viability and moisture content of the seeds when initially placed in storage, the specific Variety and condition of the storage is an environment, hence keeping seeds in a glass jar with a sealed container including desiccant in the jar, whereas the germination and viability will decline with age of seeds, viability is the ability to produce a vigorous seedling. Hence viability decline before germination, so old seeds gives weak seedlings but germinate, some standard tables are given for the perusal of the readers which are mentioned in the article johnny selected seeds (**Figures 3** and **4**) **Table 1** [4].

5. The industrial quality of seeds

The seed plants mean producing another plant thereby perpetuation the species. The safe storage will reflect the quality of the seeds and the ability of the plant to provide seeds as stores of nutrients has also made them an attractive food source for humankind, making the transition from hunter-gatherer to a settled agriculture existence and

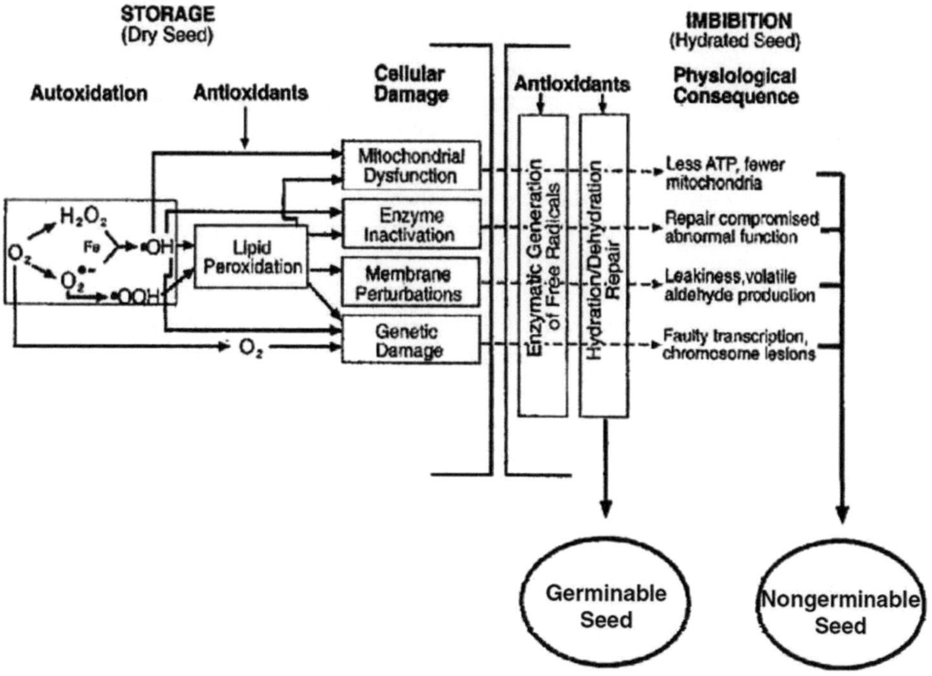

Figure 3.
A model of seed deterioration and its physiological consequences during seed storage and imbibition (source: M. B. McDonald, 1999, seed deterioration: Physiology, repair and assessment, seed science and technology 27:177–237. Reproduced with permission).

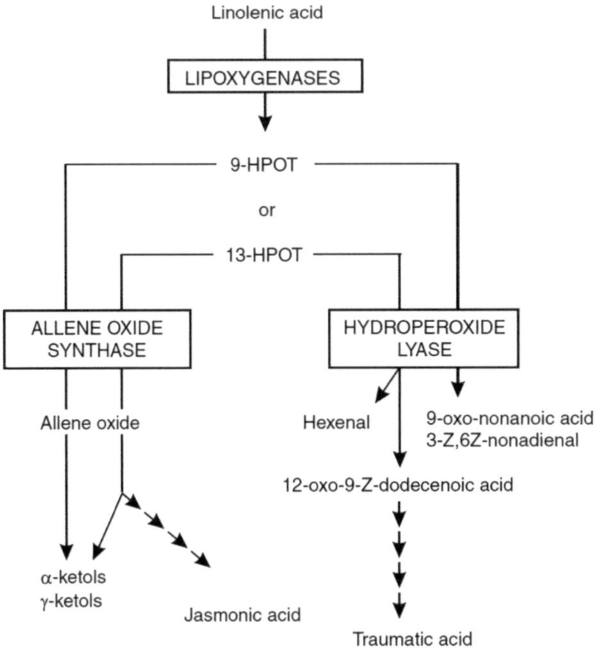

Figure 4.
Overview of the lipoxygenase pathway. 9-HPOT, 9(S)-hydroperoxytrans-10, cis-12, cis-15octadecatrienoic acid; 1-HPOT, 1(S)-hydroperoxy-cis-9, trans-11, cis-15-octadecatrienoic acid. (source: From Loiseau et al., 2001. Reprinted by permission of CABI publishing, Wallingford, Oxon, UK).

natural drug discovery process (**Table 1**) represent grain species utilized as food for man, animal and available bioactive. This development led in turn, to the building of permanent dwellings and a wide range of cultural activities. Whereas this bioactive form of dicotyledonous grains and monocotyledonous grain (cereals), will play a very important role in daily life, in some individual grains causes dietary issues, hence whole grain and fiber food inclusion will give proper nutrients vitamins and minerals because of the increased number of fibers in the diet. Whereas drugs which have food value will be more sustainable and easily binds to the target part of the cell and enzymes referee the (**Table 1**), to be concluded over here if germination to storage all the protocol followed properly and seeds and cereals are healthy and loaded with all forms of nutrients will make them become a source to get healthy bioactive out of it, to utilize as drug molecule to treat different disorders associated with the human body without involving harmful synthetic substances. The process of treatment is from "nature to creature and creature to cure with the sustainable process" [5].

6. Phytochemistry of seeds

The phytochemistry associated with the pharmacological and therapeutic effects elicited by plant material, literature survey states seeds were observed to contain appreciable amounts of alkaloids flavonoids, anthraquinone, saponin, and terpenoids and tannin (**Table 1**). As mentioned above all factors affect on the composition of chemical constituents. This chapter mainly focuses on the metabolic properties of healthy seeds and the benefits to utilize the bioactive as a drug molecules. In addition,

DOI: http://dx.doi.org/10.5772/intechopen.106208

Family	Examples	Descriptions	Secondary metabolites
Endospermic seeds (flower and fruits)			
Cucurbitaceae Core Eudicots Rosid clade	**(Cucumis melo)**	In muskmelon seed the embryo is surrounded by a perisperm/ endosperm envelope. Callose (Β-1,3-glucan deposition in this envelope is responsible for the apoplastic semi permeability of muskmelon seeds. The perisperm/endosperm envelope is weakened prior to the completion of germination	**Phenols, lipids fats carbohydrates vit. B12**
Fabaceae Core Eudicots Rosid Clade	**Fenugreek (trigonella foenum) graecumcrimson Clover (Trifolium incarnatum) Lucerne (Medicago Sativa)**	**Only some legume (Fabaceae) seeds are endospermic most legume seeds are non-endospermic**	**Phenols, lipids fats carbohydrates, fructose, cellulose**
Euphorbiaceae Core Eudicots Rosid clade	**Castor bean (Ricinus Communis)**	Castor beans seeds (Malpighiales) are a classical seed system reserve breakdown	**Phenols, lipids fats carbohydrates, galactose**
Brassicaceae Core Eudicots Rosid clade	**Garden cress (Lepidium stivum) mouse-earcress (Arabidopsis thaliana)**	Only some Brassicaceae seeds are endospermic, most Brassicaceae seeds are non-endospermic. Mature seeds have 1-2 cell layers of endosperm, while *Lepidium* of has a single endosperm cell layer. This Arabidopsis two-step germination two species exhibit, as tobacco, We. (distinct testa rupture and endosperm rupture) is a promising model system for *Lepidium* found that .(Müller et al. 2006 endosperm weakening	**Phenols, lipids fats carbohydrates**
	Cestroideae: subgroup Tobacco *Nicotiana tabacum-Nicotiana* other species petunia	Mature seeds of the Solanaceae family usually have an abundant endosperm layer. Well investigated examples **tobacco and tomato, which are model systems in are seed biology for the study of endosperm weakening and the regulation of germination by plant hormones** The Solanaceae family can **.and environmental factors** :be divided into two large subgroup	**Phenols, lipids fats carbohydrates**
Solanaceae - Core Eudicots Asterid - clade	*Petunia hybrida* **Solanoideae: subgroup** tomato *Lycopersicon esculentum* pepper *Capsicum annuum* Datura (*Datura ferox*)	**Cestroideae subgroup of Solanaceae (tobacco, :(petunia** Straight or slightly bent embryos and prismatic to **two-step germination (distinct testa,** subglobose seeds typically capsules as **(rupture and**	**Phenols, lipids fats carbohydrates, flavonoids, quercetin**

Family	Examples	Descriptions	Secondary metabolites
		endosperm rupture .fruits : (Salanoideae subgroup of Solanaceae (pepper, tomato Curved embryos and flattened, discoid seeds, no visible distinction beween testa rapture and endosperm rupture, often berries as fruits	
Rubiaceae Core - Eudicots Asterid - clade	Coffee *Coffea arabica*	The coffee embryo is enveloped by an endosperm tissue. The fully differentiated embryo lies inside an embryo cavity. The endosperm is surrounded by endocarp, which resembles a seed **Endosperm weakening of coffee is inhibited by abscisic** . coa .(and promoted by gibberellins (GA (**acid** (**ABA**	**Phenols, lipids fats carbohydrates, fats**
Oleaceae Core - Eudicots Asterid - clade	*Syringa* species	seeds is mainly imposed *Syringa* Low temperature dormancy by the mechanical resistance of the endosperm layer	**Phenols, lipids fats carbohydrates, lipids**
Asteraceae Core - Eudicots Asterid - clade	**lettuce *Lactuca sativa***	Lettuce "seeds" are actually fruits and have 2-3 cell layers of **endosperm weakening Lettuce** . endosperm below the pericarp has been demonstrated and the hormonal regulation of lettuce seed germination is similar to tobacco	**Phenols, lipids fats carbohydrates, sugars aglycon**
Apiaceae Core Eudicots - Asterid clade -	Celery *Apium graveolens*	Seeds of Apiaceae (Umbelliferae) contain relatively large amounts of living endosperm which completely surrounds a small embryo located at the micropylar has been **Endosperm breakdown** .end of the seed demonstrated and is promoted by gibberellins	**Phenols, lipids fats carbohydrates**
Ranunculaceae Basal Eudicots -	*Trollius* species	The seeds of basal angiosperms often have underdeveloped embryos that are embedded in abundant endosperm tissue. Two-step germination process with distinct testa rupture and endosperm rupture	**Phenols, lipids fats carbohydrates**
Poaceae and other monocot families Basal - Angiosperms Monocots -	wheat *Triticum aestivum* barley *Hordeum vulgare* Maize (*Zea mays*)	A typical monocot seed with endosperm is onion Alliaceae family. In the highly (*Allium cepa*) specialized cereal grains/ caryopses (wheat, barley, maize) the endosperm can be divided in the starchy endosperm (starch	**Phenols, lipids fats carbohydrates, amino acids**

Family	Examples	Descriptions	Secondary metabolites
		grains, food storage, dead cells, flour) and the aleurone	
	Onion *Allium cepa*	layer (living cell layer surrounding the starchy endosperm). The cereal .embryos are highly specialized in their structure	
Fabaceae Core - Eudicots Rosid - clade	Pea (*Pisum sativum*) garden bean *Phaseolus vulgaris* soybean (*Glycine max*)	Most species of the legume family ***Pisum*) pea** (Fabaceae) including and diverse beans have (***sativum*** non-endospermic seeds. The serve as sole food **cotyledons** storage organs as in the case of pea .During embryo. (*Pisum sativum*) development the cotyledons absorb the food reserves from the endosperm completely. In the mature seed the embryo is enclosed solely by the testa as the only seed regulation by covering layer. The ethylene of pea seed germination has been and seedling emergence studied in detail	**Phenols, lipids fats carbohydrates, fats proteins**
Brassicaceae Core - Eudicots Rosid - clade	Rape (*Brassica napus*) wild mustard (*Sinapis alba*) wild radish *Raphanus sativus*	Most species of the mustard family (Brassicaceae) including several species have non— *Brassica* **cotyledons** endospermic seeds. The serve as sole food storage organs as described for the non-endospermic. Fabaceae seeds	**Phenols, lipids fats carbohydrates cellulose**
Seeds with perisperm			
Amaranthaceae incl.) Chenopodiaceae) - Core Eudicots Caryophyllid - clade	sugar beet (*Beta vulgaris*) lambsquarters *Chenopodium album*	is diploid maternal **Perisperm** food storage tissue that originates from the nucellus. It is present in mature seeds of many **Caryophyllales** including the ((**Centrospermae** *Beta,*) Amaranthaceae among the (*Chenopodium* eudicots, but also in basal angiosperms like black pepper Piperaceae (*Piper nigrum*)	**Glucose fructose galactose, reducing and non-reducing sugar**

Table 1.
Category of seeds with phytoconstituents belonging to different families.

healthy seeds and grains are an excellent source of macromolecule and macromolecule. It is observed from the available literature that the seeds which have food value have properties to cure different diseases and its element. Moreover, the oldest sacred book composed of an ancient forms of formulation utilized as medicine from seeds (www.britannica.com/topic/Atharvaveda) are Rigveda, Ayurveda, Quran, Bible and other religious scriptures elaborated on the medicinal properties of healthy properties of seeds,

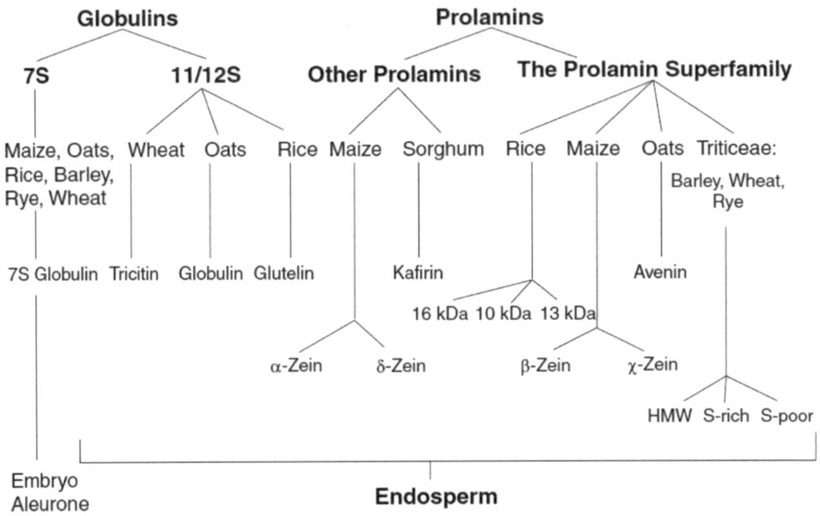

Figure 5.
Classification of the cereals, according to the types of proteins intheir grains (source: Adapted from Shewry, 1996.).

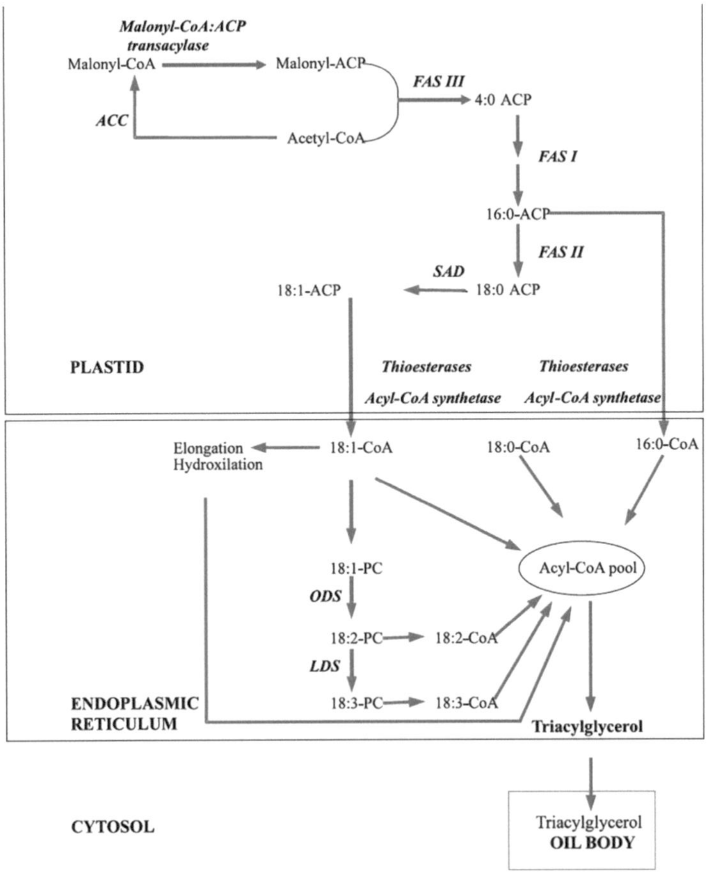

Figure 6.
Representation of triacylglycerol biosynthesis in developing seeds.

this chapter emphasizes the consequences of secondary metabolites available in healthy seeds to utilize for the cure, mitigation of different disease, disorders and its elements.

Whereas phytochemistry tie-up with agricultural science will solve the burning issues of toddy' suncontrolled viral infection or boost the immunity as well. Some important biosynthesis is given here to focus on the metabolites from a different forms of seeds (**Figures 5** and **6**).

6.1 Stages of metabolism macromolecules

The metabolic stage will begin at an early stage of germination and results in the activity of various enzymes, which are present in the dry seeds or very rapidly become active as seeds imbibe water (**Figure 7**).

6.1.1 Carbohydrates and Lipids

Sucrose biosynthesis.

6.1.2 Sucrose hydrolysis

$$\text{Sucrose} + \text{H2O} \xrightarrow{\text{Invertase}} \text{Glucose} + \text{Fructose}$$

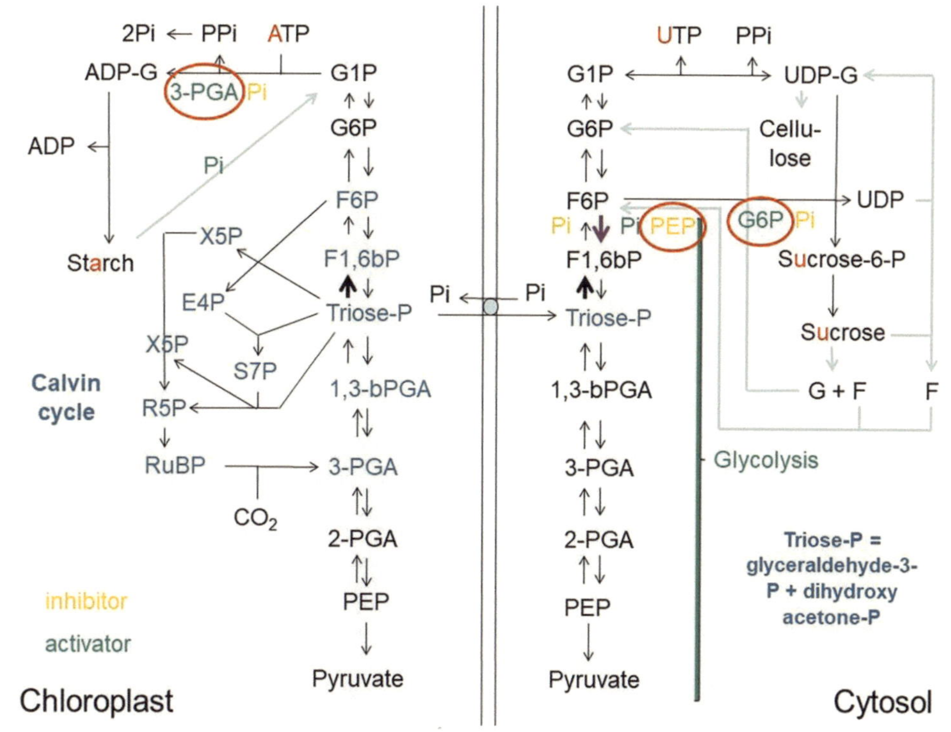

Figure 7.
*Overview of starch, sucrose and cellulose synthesis in plants and its regulatory architecture source: Google images (**Figures 8** and 9).*

6.1.3 Protein

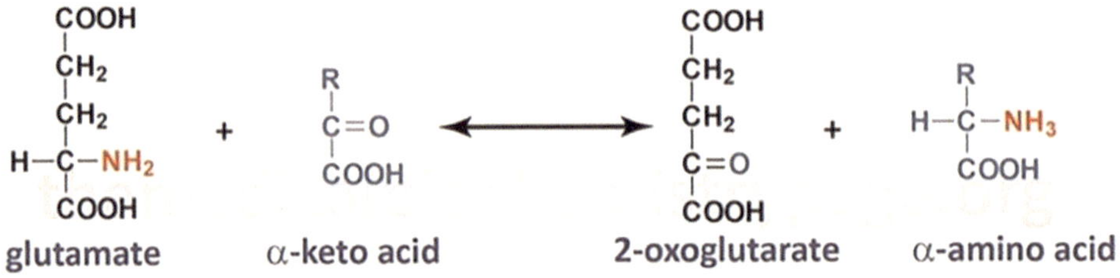

Figure 8.
(SourceRepresentative reaction catalyzed by an aminotransferase (transaminase). 2-Oxoglutarate is commonly called α-ketoglutarate. Source: Google images.

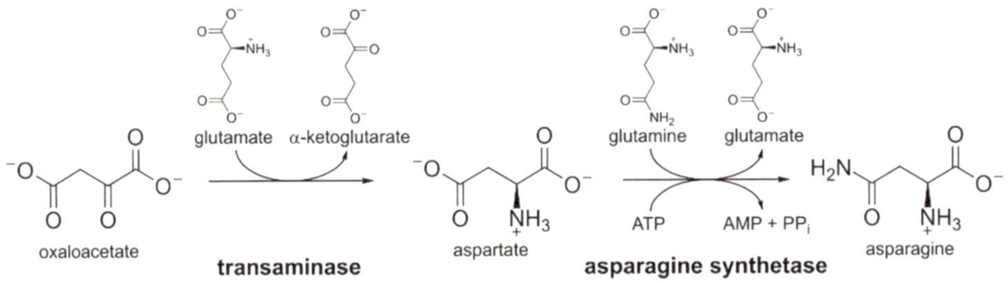

Figure 9.
The biosynthesis of aspartate and asparagine from oxaloacetate. Source: Google images.

6.1.4 Metabolisms of phosphate-containing compounds

Time of germination in days	Dry seeds	1	2	4	6
Phytin	8.6	8.5	7.2	4.0	2.0
Inorganic-P	0.4	0.3	1.9	4.8	7.0
Total lipid	0.7	0.8	0.9	0.5	0.85
Ester	0.3	0.4	0.4	0.6	0.4
RNA	0.12	0.11	0.15	0.25	0.39
DNA	0.11	0.11	0.12	0.21	0.44
Protein	0.11	0.10	0.16	0.28	0.26

Figure 10.
Metabolism of phosphate content in healthy seeds (source: Prof. Dr. Heshmat Aldesuqy biology lectures).

6.2 Metabolism of nucleic acid

The nucleic acid metabolism will give the important secondary metabolites like monomers of portions as given here (**Figures 10–15**) [5–7].

6.2.1 Synthesis of amino acid

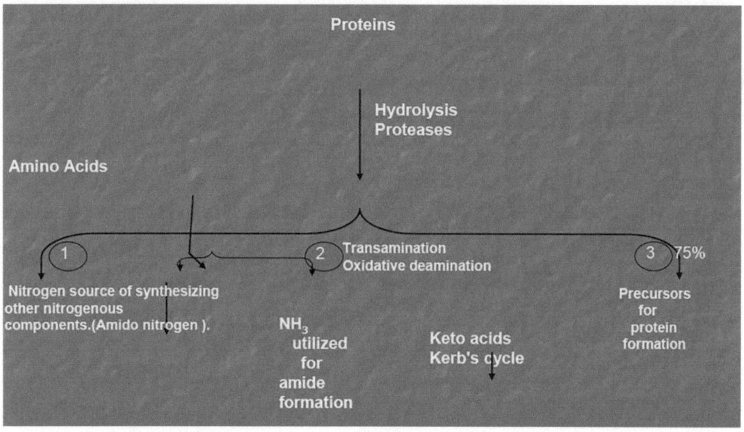

Figure 11.
Formations of monomers of amino acids source (source: Prof.Dr. Heshmat Aldesuquy biology lectures).

6.2.2 Reductive amination

Figure 12.
Formation of glutamine (source: Prof.Dr. Heshmat Aldesuquy biology lectures).

6.2.3 Trans amination

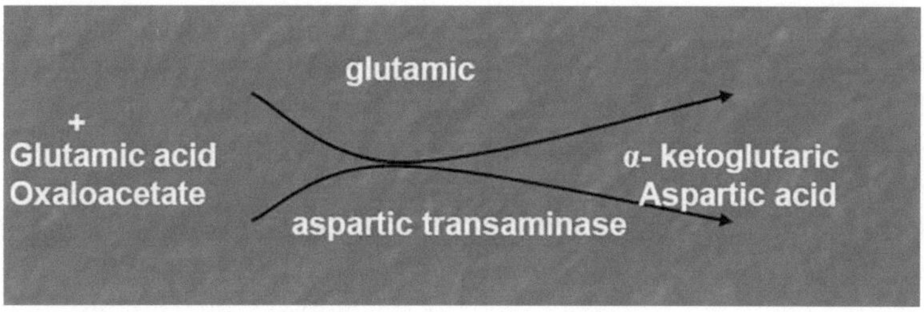

Figure 13.
Formation of aspartic acids (source: Prof.Dr. Heshmat Aldesuquy biology lectures).

6.2.4 Amidase synthesis

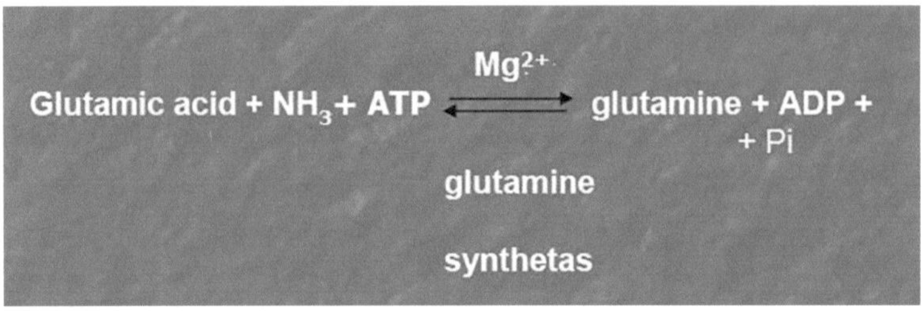

Figure 14.
Formation of glutamine (source: Prof.Dr. Heshmat Aldesuquy biology lectures).

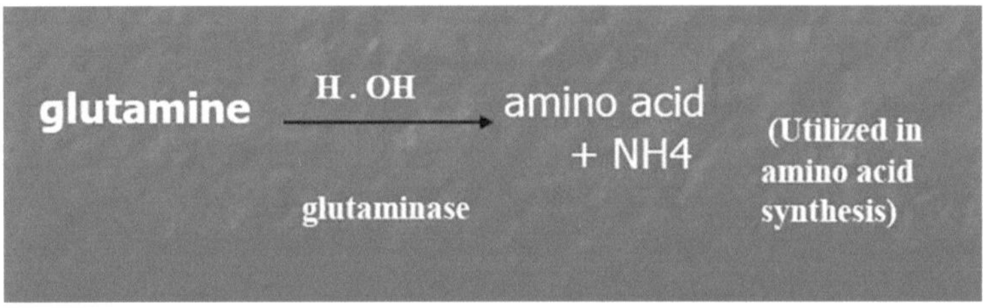

Figure 15.
(source: Prof.Dr. Heshmat Aldesuquy biology lectures).

7. Germination stimulating inhibitors affect in chemical composition

The constituents which inhibit while germination are gabrillic acid (IAA) and indoleic acetic acid (GAB3B) and kinase have been widely investigated as other pre-cursors of germination Auxins, Gibberellins, Cytokinines, Abscisic acid (ABA), Ethylene and effect of coumarin and thiourea. Factors affecting germination viability and life span, external factors affecting life span, stage of germination includes phase activation, digestion and translocation. Epicotyl and hypocotyl germination. This is the stage where seeds can be protected from denaturation and fermentation [8]. This is as discussed in the chapter at the beginning.

Tables: Seed biology with important phytoconstituents [9, 10].

8. Conclusions and future study

1. The remarkable development and environmental factors represents the qulity of seed dormancy and germination. The review explaines the indepth process of seedlings transition phase from development to metabolic growth of bioactives in healthy seeds.

2. Furthermore the hormone ABA is essential for development of seeds until the condition become favorable. This background analysis not only put forward the metabolic process in healthy developed seeds, but also helps agriculturist & phytochemist to choose an healthy seeds to isolate vigor repository in nutshell.

3. The well evalvated seed will be the perfect repository of secondry metabolites which helps to Maintaining of healthy lifestyle by combate with the disorder and disease associated with the leaving species.

4. Whatever this seeds floting on water, blown with air, carried away by animal or choosen by scientist they are the healthy babys (Seeds) scattered to expanding the geography and phytochemistry, thus avoiding the competation with parent plant.

5. To be conclusion the highly ranked seeds eliminate water requirement during sexual reproduction and allow fertilization events to occurs over long distance, where as the seeds of different plant needs different pretreatment to get vigor seedling for production of healthy crops even to isolation the healthy biomolecules.

6. We endup by saying real attention to be focused on plant propagation, particularly indigenous tree species and seedling multiplication should be our preamble to make sustainable environment. if environment is sustainable the life science research will flowrished in right way.

9. Future aspects

The regulatory issues facing both protection and maintenance, will make industry ready seeds, which will be healthy foods with proper nutrients. Hence it was a great deal to focus on seed production, protection and maintenance. The phytochemistry play an impotent role to isolate the bioactive from healthy seeds based on their metabolic pathway able to isolate different macromolecules and micromolecules which can be converted into drug molecule with food value which do not produce any kinds of adverse effect on the human physiological system,

Yes seed refuses to die when it get bury it become tree, when dirt through on it will increased its value, when seed wants to rise it dropes everything that is weighing it down. When crushed it will be the property of medicine. Grower-friendly, crop-friendly, environment-friendly and ultimately Drug molecule-friendly.

Acknowledgements

Author immensely grateful to the management of YPCRC, Yenepoya (Deemed to be University) and the book publishing family for allowing me to become a part of this novelty work. Last but not least my heartfelt thanks to the Dean of YPCRC's guidance support. I never deny the support and love of my children and family for providing me the quality time to draft this chapter.

Conflict of interest

The author declared no conflict of interest.

Notes/thanks/other declarations

Thanks to the publishing house for allowing me to contribute.

Acronyms and abbreviations

ACC	acetyl-CoA carboxylase
FAS	fatty acid synthetase
SAD	stearoyl-ACP desaturase
ODS	oleoyl-phosphatidylcholine desaturase
LDS	Linoleoyl-Phosphatidylcholine desaturase
PC	Phosphatidylcholine
ATP	Adenosine triphosphate
ABA	Abscisic acid

Author details

Nikhat Farhana[1*], Ripudaman M. Singh[2], Mohammed Gulzar Ahmed[1], Thouheed Ansari[3], Abdul Rahamanulla[1], Ayesha Sultana[1], Treesa P. Varghese[1], Ashwini Somayaji[5] and Abdullah Khan[4]

1 Department of Pharmaceutical Chemistry, Yenepoya Pharmacy College and Research Center, Yenepoya (Deemed To Be University) Mangalore, India

2 UIPS, Chandigarh University, Gharuan, Punjab, India

3 Dr. Noor Mohammed Khan General Hospital, Hafr Al Batin, KSA

4 Centre of Excellence for Pharmaceutical Sciences (CoEPS), School of Pharmacy, KPJ Healthcare University College, Nilai, Malaysia

5 Nitte University, Mangaluru, India

*Address all correspondence to: nik.nida13ada25@gmail.com

References

[1] Nonogaki H. Seed biology updates – Highlights and new discoveries in seed dormancy and. Germination. 2017;**8** (April):1-16

[2] Cbf IOF. Seed biology – From lab to field. Journal of Experimental Botany. 2017;**68**(4):761-763

[3] Nonogaki H. Seed biology updates – highlights and new discoveries in seed dormancy and germination research. Frontiers in Plant Science. 2017;**11**:1-16

[4] Ashwlayan VD, Antlash C, Imran M, Asdaq SMB. Insight into the biological impact of COVID-19 and its vaccines on human health. Saudi Journal of Biological Science. 2022;**29**(5):3326-3337

[5] Hern A. Seed science research: Global trends in seed biology and technology. Seeds. 2022;**1**:1-4

[6] Bareke T. Biology of seed development and germination physiology. Advances in Plants and Agricultural Research. 2018;**8**(4): 335-346

[7] Benech-arnold RL, Sánchez RA. Handbook of Seed Physiology. New York – London: The Haworth Press; 2005

[8] Ave B, MEP F. Seed Storage Guide. Vol. 955. Johnny's Selected Seeds; 2021. pp. 1-5

[9] Duval M, Gallardo K, Catusse J, Bally J, Job C, Job D. Seed germination and vigor. Annual Review of Plant Biology. 2012;**63**:507-533

[10] Adeosun AM, Oni SO, Ighodaro OM, Durosinlorun OH, Oyedele OM. Phytochemical, minerals and free radical scavenging profiles of Phoenix dactilyfera L. seed extract. Journal of Taibah University Medical Sciences 2016;**11**(1):1–6. Available from: DOI: 10.1016/j.jtumed.2015.11.006

Chapter 3

Seed Dormancy: Induction, Maintenance and Seed Technology Approaches to Break Dormancy

Tabi Kingsley Mbi, Ntsomboh Godswill Ntsefong and Tatah Eugene Lenzemo

Abstract

Dormancy is the major cause of erratic germination, patchy emergence and uneven seedling establishment in the field. These traits are exceedingly undesirable in crop production as future phases of growth and development are strongly linked to uniform seedling development at early growth phases. Variations in maturation time, and difficulty in managing abiotic and biotic stresses during pre- and postharvest are common consequences of uneven germination and seedling emergence. Minimizing this negative impact of dormancy in a seed lot is the major concern of all seed production companies. Generally, mature seeds show some considerable dormancy during which embryo growth is halted momentarily because one or more internal and external stimuli for growth resumption is/are absent. If the inhibition of seed germination is solely due to insufficient or complete absence of external signals, then the seed is in a state of quiescence. Otherwise, if linked to internal factors, then the seed is in a state of dormancy. Induction, maintenance, and release of dormancy are therefore related to Seed-dependent factors such as morphology, hormones, state of embryo maturity at seed dispersal and chemical inhibitors. This chapter focuses on species-dependent methods currently used to break dormancy, reduce germination time and improve emergence and seedling establishment.

Keywords: seeds, germination inhibitors, dormancy, scarification, embryo rescue

1. Introduction

Generally, the concept of dormancy is centered on the absence of growth in any plant organ having a meristem like bulbs, corms, axillary buds, seeds and even in other living organisms like fungi spore, spirogyra zygospore etc. In the discipline of seed biology, dormancy could be considered simply as a block to the completion of germination of an intact viable seed under favorable conditions. The process might look simple but several authors have reported that dormancy is one of the least understood phenomena in the field of seed biology and needs further elucidation [1, 2]. The resting condition of many seeds, especially those of different grasses and garden crops, is maintained only as long as the seeds are in dry storage. As soon as a suitable moist medium and a favorable temperature are provided germination proceed almost immediately. Dormancy is a seed characteristic, manifesting as a block or series of blocks that prevent germination under otherwise favorable moisture, temperature and gaseous conditions. Dormancy is thus considered as an adaptive life history trait to seasonally unfavorable environmental conditions [3]. Dormancy is just one among quite a wide range of reasons why a seed may not germinate [4]. The concept of dormancy signifies that the miniature plant has life but require other factors beyond external factors like temperature and water to resume growth. Inhibition of germination is therefore completely or partially linked to a combination of three seed borne mechanisms, notably Chemical inhibitors that prevent growth, physical barriers that prevent the uptake of water, gases or chemicals and incomplete development of embryo prior to seed dispersal. In the later mechanism, the embryo of the seed needs extra time after dispersal to ripen [5, 6], and optimum levels of internal hormones [7]. Seed germination is the most critical part in the life cycle of seed bearing plants and seed dormancy is an excellent capability to increase the chance of survival by optimizing the distribution of germination in time or space [8].

The significance of seed dormancy can be seen from the ecological and agronomic stand points. Ecologically, seed dormancy can be beneficial for propagation and dissemination of plant populations while in agronomic systems, dormancy is a problem for seed evaluation and seedling establishment. For example it has been reported that when seed dormancy level is too low, it can trigger pre harvesting sprouting leading to either yield loses in cereals or germination before the start of a favorable growth season, risking seedling mortality. In contrast, deep seed dormancy levels limiting production in many field crops as it turn to delay germination and reduce the length of the growing season [3, 9].

1.1 Definition

There is yet no consensual definition for dormancy therefore many definitions and classifications exist [1]. It is a situation whereby a viable seed fails to germinate though given presumably favorable conditions [10]. Dormancy in a strict sense will refer to the inability of an intact viable seed to resume growth even when the environmental conditions are most favorable [1, 11]. Dormancy is delay of a viable seed to germinate, under a given set of environmental conditions and will only germinate if restricting factors are eliminated either via natural or artificial means [12]. Based on the later definition, it is evident that the state of dormancy can only be experimentally proven when the seed is placed under suitable external growth conditions and nonetheless germination process is not initiated, even though embryo is a life. The process is thus complex as it is difficult to appreciate the degree of restriction imposed by each seed borne characteristics contributing to delay resumption of growth by the embryo.

1.2 Significance of dormancy

Whether a seed should remain dormant or proceed to germination under certain circumstances is important in two aspects; first with respect to its survival as a species under specific ecological conditions and secondly relating to its economic and agronomic importance. From an ecological viewpoint, dormancy is an important survival mechanism that favors propagation and dissemination of seeds to establish plant populations. Because specific conditions are required to break dormancy, it may favor germination and seedling emergence under more favorable conditions [2].

Heterogeneous and asynchronous seed germination in the soil over years due to dormancy is an extra advantage to the survival of some species. This is because individual seeds in a seed population usually have different levels of dormancy, spreading their germination over time. This delay avoids unfavorable environmental events such as a drought that would eliminate the population if all the seeds were to germinate at same time on one hand and intra-specific competition for the available resources within the ecosystem on the other [13]. This explains why weeds are difficult to eliminate in a field because the seed banks provide a vast array of seeds with differing levels of dormancy. Although deep seed dormancy is considered problematic in agricultural species, some level of dormancy is desirable to prevent vivipary, a phenomenon rampant in cereals in which pre-harvest sprouting occurs. Vivipary causes losses in seed quality and quantity in agricultural plants. Regardless of all the advantages of dormancy for natural plant populations, it is an undesirable trait for most economic crops because it makes rapid and uniform germination during crop establishment difficult.

2. Key factors in the induction/maintenance of dormancy and initiation of germination

2.1 Environmental factors

The stimuli needed for seed germination include chemical or hormonal signals or environmental signals such as temperature, soil nitrate levels, or light [1]. It is important to keep in mind when dealing with dormancy and germination that a seed is never just under the control of one factor in nature, but many factors concurrently [5].

2.1.1 Temperature

Seed germination is a complex process involving many individual reactions and phases, each of which is affected by temperature. Because temperature influences both the percentage and rate of germination of seeds, it is one of the most critical factors affecting seed germination. Although seeds of each species have optimal temperatures for attaining maximum germination, between 30 and 40°C for most palms [14]. Alternating temperatures are preferred to constant temperatures because they can overcome shallow seed dormancy and enhance uniform germination. In the wild, most tropical tree seeds generally require higher temperatures to germinate; for example Soil temperatures above 38°C, but below 42°C can reduce the time required for germination of seeds of *E. guineensis* from years to weeks [15, 16].

2.1.2 Water

Water is a basic requirement for germination and its role as a medium for biochemical processes leading to germination, such as weakening the seed coat, activating enzymes, and breaking down food reserves, scarcely requires emphasis. Physical dormancy has been attributed to a hard or water-impermeable seed cover [17], such as a fibrous mesocarp and/or a stony endocarp, which are very common among palms [18]. It is generally recognized that seed germination is more sensitive to moisture stress than its subsequent seedling development [19]. Though moisture favors germination, excess of it may also become an obstacle to germination due to improper ventilation for the physiological process of germination [18]. While inadequate moisture results in delayed and poor germination, excessive moisture will hinder germination due to decreased aeration. Metabolic activation, preparation for cell elongation, radicle emergence, and seedling growth are subsequent events of seed germination and require different levels of hydration [20].

2.1.3 Oxygen

Seeds of many species will not germinate well at an oxygen level considerably lower than that normally present in the atmosphere [19]. From the physiological point of view studies are still to prove that inadequate oxygen supply induces dormancy [21]. Nevertheless, it has been acknowledged that oxygen is required to break chemical dormancy caused by germination restriction substances located in the endocarp [22]. However, insufficient oxygen supply under excessive moisture in a growth medium delays initiation of germination [23, 24]. In the tropics oxygen availability is usually not a limiting factor since germination usually takes place at room temperatures [25]. It has been reported that germination of oil-palm seed is dependent upon a minimal threshold concentration of oxygen in the embryo [22].

2.1.4 Light

The promotional effect of light in seed germination is through a single photoreaction controlled by the blue pigment phytochrome [26]. Seeds of many temperate tree species are known to be light sensitive, and their germination is promoted by red light and inhibited by far red light [27]. It is generally held that both light intensity (lux) and light quality 'colour and wavelength' influence seed germination. Germination is triggered by increases in light as well as by the ratio of red to far-red light and temperature [28]. However, while light has been reported to be a germination inhibitor i.e., show negative photoblastism in some Palmae as it's the case of *Sabal palmetto* [29] others are indifferent to light for their germination. Germination of seeds covered by soil suggests that seed of most species are indifferent to light [30].

2.1.5 Soil factor

Edaphic factors like soil structure, texture, humidity, pH and temperature among others influence seed germination and emergence.

2.2 Internal factors

2.2.1 Seed maturity

Seed maturation is the second phase of seed development after embryogenesis during which food reserves accumulate, while dormancy and desiccation tolerance develops [31]. In angiosperms, there is increasing evidence that maternal environmental effects affect both germination percentage and germination phenology [32]. Environmental factors that affect the mother plant, such as photoperiod and temperature, during seed maturation have also been found to influence seed germination of many species [33]. These factors may also influence the size of the seeds, which in turn may influence germination timing and success [34], and within species, seed mass is often associated with probability or time of germination [35]. Identifying the environmental mechanisms involved in the transmission of maternal environmental effects through the seeds is important to understand their ecological function and adaptive value [32].

2.2.2 Plant growth regulator (PGR)

ABA, GA, ethylene, brassinosteroids, auxins, and cytokinins have a tremendous effect on plant development, even at low concentrations among all PGR [36]. Even though PGR like cytokinin, auxins ethylene and brassinosteroids can either promote germination alone or in association with ABA, most literature on hormonal regulation of seed germination suggests that induction and release of dormancy is controlled by two main plant growth regulators notably ABA and GA. However, the influence of these two hormones is mainly at the variation in their ratio than their absolute content. Thus, dormancy induction would depend on ABA and GA metabolism during the process. Any metabolic processes that can change the concentration of the active forms of IBA and GA within the seed like synthesis, degradation, activation or deactivation will either induce or break dormancy. Thus any metabolic process that synthesizes or activates IBA induces dormancy because such processes turn to degrade or deactivate GA, given that these two hormones are antagonistic. Increase of seed to ABA sensitivity, and decrease of GA sensitivity induce dormancy and the opposite events will lead to dormancy alleviation and germination [37].

2.2.2.1 Abscissic acid

ABA is a sesquiterpene hormone that plays an integral part in mediating the adaptation of the plant to biotic and abiotic environmental stresses, regulation of seed development and germination. During seed development, ABA concentration increases intensely during embryogenesis to promote accumulation of reserves, prevent precocious germination, promote embryo tolerance during seed desiccation phase, and induction of dormancy at dispersal [38]. ABA is a main plant growth regulator of dormancy induction and maintenance [1]. Insensitivity to ABA during seed development results to precocious seed germination. This has been observed in maize, tomato, and Arabidopsis mutants deficient in ABA content [39, 40].

In spite of the fact that the relationship between ABA content and its physiological function in seed dormancy and germination is still to be made clear, more light has been shed on mechanisms surrounding ABA-mediated seed germination inhibition. It is the case in *Brassica napus*, where it was suggested that one of the probable contribution of ABA to inhibiting germination is to prevent loosening of the embryo cell wall which impedes water uptake [41]. In addition, ABA has also been found to specifically inhibit endosperm rupture instead of testa rupture. The effect of ABA on seed dormancy can be efficiently alleviated by chilling 'stratification treatment' so that endogenous ABA content drops precipitously with a concurrent increase in germination rate [42]. This inhibitory effect can be partially counteracted by the antagonistic action of GA [43].

2.2.2.2 Gibberellic acid (GA_3)

Gibberellins are a group of tetracyclic diterpenoids endogenous regulators that are well-known for their capability to promote plant growth and development [44]. In the process of seed development, GA levels are usually high during the embryo morphogenesis phase and decreased during the embryo maturation phase [45]. Active GAs may help promote the growth of embryo and later they are deactivated to avoid vivipary [46]. There are reports on the application of GA_3 in alleviating innate and environment-induced dormancy [7, 47] or synchronizing GA with different scarification pre-treatments like stratification, heat and removal of seed coat [5, 48]. In fact, GA is likely to antagonize the ABA effect on seed development, particularly dormancy induction [49].

2.2.2.3 Mechanism of GA3 in promoting germination

The role of GA3 in breaking of dormancy has been documented by different workers [50, 51]. The hormone signal (GA_3) activates DNA in the aleurone cells. This is followed by transcription and translation of a gene coding for α-amylase and its eventual production inside the aleurone cells. In the model for gibberellin-induced production of α-amylase, it is demonstrated that GAs produced in the scutellum diffuse to the aleurone cells, where they stimulate the secretion of α-amylase [52].

2.2.2.4 Regulation of dormancy/germination by GA and ABA

The two main plant growth regulators that regulate dormancy-germination cycling are ABA and GA_3. A model for the regulation of dormancy by ABA and GA in response to the environment has been proposed [37]. This model suggests that, the sensitivity of embryo to ABA and GA ratios within the seed is influenced by environmental factors like temperature. If the ABA synthesis and signaling (GA catabolism) dominates then dormant state is induced and maintained while the opposite event promotes transition to germination. Variations in the level of dormancy alter the requirements for germination when these overlap with varying environmental factors. Throughout the process of seed development, dormancy can be induced and maintained depending on GA/ABA balance. During imbibition, GA levels increases while ABA levels decline, suggesting that GA and ABA have antagonistic roles in seed dormancy and germination processes [53]. GA counteracts the effect of ABA by promoting the embryo growth potential and the weakening of tissues covering the embryo [54].

2.3 Multiple control points in dormancy induction and breaking

A block to germination may occur at more than one point in the sequence of steps leading to visible growth [55]. Various dormancy mechanisms occur at different levels in the course of seed development and principally in germination, indicating the concurrent nature of dormancy and germination. For example a dormant seed without a hard seed coat will imbibe water but remain blocked at some stage of metabolic activation or growth until other blocks are removed [56]. For example an impermeable seed coat inhibits water absorption, blocking imbibition, and, in many cases, the seed coat may reduce oxygen availability to the embryo [57]. If a seed successfully completes imbibition, it will normally be metabolically active [1]. However, germination-related activities may further be blocked during activation phase in dormant seeds. Even if this stage is completed successfully, a mechanical restriction of root growth may still impose dormancy on the seed, at the last step in germination. This shows that dormancy induction can occur in more than one control point depending on the plant species. The complex interactions between stimulating, inhibiting and limiting factors demonstrate that a series of requisites must be met for the seed to germinate. A strong line of evidence in support of multiple control points in dormancy induction and breaking is provided by [58] who characterized germination of *Sisymbrium* and *Arabidopsis* seeds under different wave lengths of the visible spectrum and growth stimulants.

Secondary dormancy may be induced by metabolic variations, like ABA synthesis due to drought or cold ambient conditions or when the level of primary dormancy decrease, during after-ripening, because of decreases in ABA level or increases in GA or cytokinin levels [59]. At the molecular level, a well-known genetic barrier to germination is DELAY OF GERMINATION1 (DOG1) [60]. Though the binding of (DOG1) to Protein Phosphatase 2C ABSCISIC ACID (PP2C ABA), Hypersensitive Germination (AHG1) and heme are independent processes, they are however both essential for the *in vivo* functioning of DOG1's. AHG1 and DOG1 are both involved in the regulatory system for dormancy induction, maintenance and germination. DOG1 has a significant influence on the level of expression of ABA INSENSITIVE5 (ABI5) [61, 62].

3. Classification of dormancy

Generally, different classes of dormancy require different dormancy breaking methods [4]. By misinterpreting the class of seed dormancy, researchers may fail to apply the correct methods to overcome dormancy in a given species. For this reason, method of overcoming dormancy must be directed towards the specific kind of dormancy [63]. Owing to the fact that the notion of dormancy still need to be elucidated, several classifications exist, but the most common are those proposed by Nicolaeva [11], Hilhorst [2] and that of Baskin and Baskin [64].

3.1 Classification based on barrier factors

One of the earliest classifications of dormancy is that proposed by [11] who distinguished three main types of dormancy based on barrier factors within the seed.

3.1.1 Exogenous dormancy

Dormancy is due to some features of the seed located outside the embryo. Exogenous dormancy could be due to the following; impermeability of seed coat to water owing to seed coat structure, which is hard enough to restrict the entry of moisture into the seeds. Secondly seed coat may be impermeability to gases, hence insufficient intake of oxygen necessary for catabolism of reserves. Thirdly, extremely hard seed/fruit structure such as seed coat, endosperm as in Acacia species can impose mechanical resistances on further development of the embryo. Finally, the testa may produce some biochemical substances that serve as inhibitors, blocking germination of embryo.

3.1.2 Endogenous dormancy

The cause of this dormancy is present within the embryo and can be as a result of; incomplete embryo development after ripening period, in such cases, germination does not occur until the embryos develop to their normal size. Common in family

Palmaceae, Amgnoliaceae; inhibitors are present within the embryo. In such cases germination can commence only when these inhibitors are leached out of the embryo, this is the case of Xanthium, Fraximus.

3.1.3 Combined dormancy

This type is provoked by a combination of two or more exogenous and endogenous factors which act in complementary fashion.

3.2 Classification based on time of induction

This classification suggested by [2] distinguishes only two types: primary and secondary dormancy.

3.2.1 Primary dormancy

Primary dormancy is induced during the seed maturation phase and reaches a high level in freshly harvested seeds, meaning seeds with primary dormancy are dispersed from the mother plant in a dormant state [1]. During subsequent dry storage of seeds (after-ripening), dormancy slowly reduces [65]. Primary dormancy is maintained by the accumulation of abscisic acid (ABA) during seed maturation to prevent viviparity [66] and requires a period of after-ripening before seeds have the capacity to germinate under favorable conditions. The level of primary dormancy in seeds is determined by several factors of genetic and non-genetic origin [67]. All these factors may cause physiological variability which is matched with differences in seed morphology (size, weight, color etc.) or simply heterogeneity in degree of dormancy [5]. Developing seeds rarely germinate, and when precocious germination does occur, it is frequently associated with deficiencies in ABA synthesis or sensitivity [2]. The induction of primary dormancy is linked to the two peaks of ABA accumulation during developmental phases of seeds. As observed in studies carried out on Arabidopsis, the first ABA peak occurs prior to embryo maturation hence it's described as maternal. The second peak regulated by the genome of the embryo is observed during maturation hence described as embryonic ABA peak. The accumulation of embryonic ABA, but not maternal ABA, is indispensable for the induction of primary dormancy [68].

Primary dormancy has the advantage of guaranteeing complete development of embryo, and prevents precocious germination in species like maize. Once seeds acquire primary dormancy during late maturation its water content decreases drastically. It is the case with some *Arabidopsis* accession Cvi, in which water content decreased by almost 5-fold was noticed after acquiring primary dormancy to late maturation phase [69]. When the moisture content is further reduced to a certain level by dry storage, the seed loses primary dormancy. This process of breaking primary dormancy includes a decrease in ABA concentration and sensitivity, an increase in GA and light sensitivity, and a widening of temperature range for seed germination [70].

3.2.2 Secondary dormancy

Secondary dormancy is generally induced by unfavorable environmental conditions following shedding of mature seeds from the parent plant which were not dormant at shedding [2]. Secondary dormancy is induced when changing environmental conditions cause undesirable germination conditions, such as unfavorable temperature, extended light or darkness, water stress, or lack of oxygen. This type of dormancy does not only decrease with time, but it can also be re-induced in non-dormant seeds when conditions for germination like light are lacking [26]. When exposed to inappropriate conditions like critical temperature, anoxia, limited light etc. secondary dormancy might occur after imbibed after-ripening seeds have lost primary dormancy. In the soil seed bank, secondary dormancy enables cycling, through which different depths of dormancy are progressively gained or lost, until the environment is favorable for germination, and then seedling establishment [6].

3.3 Classification according to Baskin and Baskin

Baskin and Baskin [64] based on an initial dormancy classification scheme proposed by Nikolaeva [11] to put forward a more comprehensive hierarchical classification system made up of five classes of seed dormancy which are in turn sub divided into levels and types.

3.3.1 Physiological dormancy (PD) or class A

Physiological dormancy is present among species distributed over the entire phylogenetic tree of gymnosperms, basal angiosperms, monocots and eudicots [64, 70]. PD is caused by low growth potential of embryo, which cannot overcome mechanical constraint of seed (or fruit) coat. It is the most abundant form of dormancy common in seeds of angiosperms and all major angiosperm clades. Class A has three levels: deep, intermediate and non-deep.

3.3.1.1 PD deep

Embryos excised from seeds with deep PD either do not grow or produce abnormal seedlings. GA treatment does not break the level of PD dormancy. Three to four months of cold (subtype a) or warm (subtype b) stratification are needed before germination can occur. Examples of plant family of sub type are *Aceraceae*, and b is *Ericaceae* [71].

3.3.1.2 PD intermediate

Embryos excised from such seeds produce normal seedlings. GA treatment promotes germination in some but not in all species. Seeds require 2–3 months of cold stratification. Dry storage after-ripening can shorten the cold stratification period. Example of a plant with PD intermediate is Acer Pseudoplatanus (*Aceraceae*) [70].

3.3.1.3 PD non-deep

Embryos excised from such seeds produce normal seedlings. GA treatment can break this dormancy but depending on the dormancy can also be overcome by scarification, after-ripening in dry storage, and cold (0–10°C) or warm (>15°C) stratification. The sensitivity of seeds to light and GA increases as non-deep PD is released. Members of Asteraceae and poaceae have non-deep PD [72].

3.3.2 Morphological dormancy (MD) or class B

Class B dormancy is evident in seeds whose embryos are under developed in terms of size but well differentiated cotyledon and hypocotyl-radical. The embryos are not physiologically dormant but simply needs time to grow before seed germinates, i.e., growth period of embryo = period of dormancy.

3.3.3 Morpho-physiological dormancy (MPD) or class C

It is common in seeds that have underdeveloped embryos 'in terms of size', supplemented with physiological component to their dormancy. These seeds therefore need a dormancy-breaking treatment that will take in to account both morphological and physiological constraints. For example a defined combination of warm and/or cold stratification which in some cases can be replaced by GA treatment. In MPD-seed, embryo growth/emergence requires a considerable longer period of time than in MD-seeds.

3.3.4 Physical dormancy (PY) or class D

PY is known to occur only in angiosperms [17]. Seeds with physical dormancy possess water impermeable fruit coats 'pericarps' or seed coats 'testa'. Impermeability might be enhanced by many factors among which are several layers of tightly spaced, thick-walled cells in the pericarp and testa, the presence of waxes, lignins, and pectins, or a combination of these factors, all impeding water uptake during imbibition [17]. As far as these impermeable layers remain intact, dormancy is maintained until when subjected to natural or artificial conditions that will disintegrate the impervious layers. Wide temperature variations, fire, drying, and passage through gut of animals are among natural factors that can disintegrate impervious layers of the pericarp. Mechanical or chemical scarification is common practices used in seed technology to break PY dormancy. In several cases, specialized structures are associated with the control of water-impermeability, for example, in PY-dormant *Anacardiaceae* species (Rosids), the endocarp consist of three water impermeable palisade layers (macrosclereids, osteosclereid and brachysclereids) and the outer crystalliferous layer which block water entrance to the carpellary micropyle in dormant seeds [73]. Once the impervious layers becomes permeable following natural or artificial scarification methods, the later can no longer return to their initial stage of absolute impermeability. Thus the timing natural release of dormancy is a more significant in the life cycle of plants with PY, than it is in plants with PD [74].

3.3.5 Combinational dormancy (PY + PD) of class E

Class C dormancy is found in seeds with water impermeable seed or fruit coat (PY) synchronized with physiological embryo dormancy (PD none deep). Release from PY and PD of PY + PD-dormant seeds appears to be independent events and the timely order can be species specific [64].

4. Different methods of breaking dormancy in seed technology

4.1 Scarification

Scarification is a technique used to break mechanical dormancy [75] which acts by impeding the perception of germination elicitors like water, light, temperature and oxygen by the embryo. Scarification techniques used commonly include; mechanical, thermal, hot water and chemical scarification. This type of dormancy breaking has been documented in a number of families e.g. *Anacardiaceae*, *Arecaceae*, *Cornaceae*, *Elaeagnaceae*, *Empeteaceae*, *Juglandaceae*, *Meliaceae*, *Nyssaceae*, *Oleaceae*, *Rhamnaceae*, *Rosaceae* and *Santalaceae* [76].

4.1.1 Mechanical scarification

This method can be accomplished, by abrading the surface of the seed until the endosperm becomes visible, or by using a knife to scrap out the hair plug at the micropylar opening. This method gives the possibilities to bring the best of germination capacity of seeds, although its application is difficult for significant quantities of seeds. Mechanical scarification accelerates germination and some chemical treatments significantly increased germination speed of the mechanically scarified seeds [77].

4.1.2 Thermal scarification

Thermal scarification can be by dry heat or hot water on the other hand can be realized by subjecting seeds under a high and steady temperature of a given duration. In the case of oil palm, dormancy is only broken when seeds are exposed to a constant thermal scarification temperature of 40°C for 80 days [78].

4.1.3 Hot water treatment

This treatment involves soaking seeds in water at 40–100°C depending on the species and seed coat thickness, for a specific period of time or until the boiling water cools to room temperature [79]. A brief soak in 80°C water for 10 minute resulted to 91.26% germination of *Acacia catechu* while soaking in 100°C water for a period of 12 min for *Elaeocarpus floribundus* gave 84% seed germination success rates [80]. Improvement of germination via soaking in hot water could be associated to weakening of seed coat. Such weakening probably occurs because lignins and pectins present on epidermal layer of the seed coat are dissolved, hence water and oxygen signals are perceived by the embryo [81].

4.1.4 Acid scarification

This method is recommended only for those seeds that are very hard to germinate, as damage to the embryo during the process can be high [82]. The treatment generally requires soaking seeds in 95% pure (1.84 specific gravity) sulfuric acid. The soaking duration is a factor of the degree of thickness of the pericarp of a given species. Once the soaking time elapses, acid is decanted, then seeds washed severally and dried [83]. The timing of this treatment is critical therefore the soaking period and the post soak washing have to be precisely controlled to avoid seed injury. The acid scarification can be applied either at room temperature or in a heated condition by soaking the seed in different concentrations of sulfuric acid for 10–30 min [79].

4.2 Alternating soaking and drying in water

This is the simplest treatment to give the seeds an early start in the germination process. It is also known as invigoration and its effects are not only on the activation of enzymes and mobilization of reserves in the aleorone layer [84], but also on the softening of hard seed coats and leaching out of chemical inhibitors. Aerated, cold-water soaking for 28 days at 11°C was found to be effective in breaking moderate dormancy and enhancing germination of *Pinus taeda* seeds [85]. Soaking and drying treatments can have varying effects on germination depending on the rate of drying, the species tested, and the duration of the soaking [86].

4.3 Seed stratification

Stratification is used mainly to alleviate dormancy in temperate plant species. The stratification approach depends on the cause of dormancy in a species. For example dormancy due to immature embryos is broken via warm stratification; physiological dormancy is relieved by cold stratification while combined warm and cold stratification is effective for seeds that have both immature embryos and physiological dormancy [76].

4.4 Embryo rescue

Embryo rescue is a dormancy alleviation method in which immature or mature zygotic embryos are isolated and cultured under aseptic conditions on an aseptic nutrient medium [87]. While the culture of immature embryos has as objective to prevent embryo abortion or sudden arrest of growth during ontogeny, one of the reasons the mature embryos are cultured is to eliminate seed germination inhibitors or to shorten the breeding cycle if dormancy is a problem [88]. The factors which cause dormancy are endogenous inhibitors and embryo immaturity. Breeding cycle can be shortened by removing the embryos from the influences of these factors which localized in the seed coat and endosperm, or both. The breeding cycle of papaya could be shortened via embryo culture from 6 to 9 months to approximately 3 months [89].

5. Conclusion

In this chapter, the definition, significance, classification, key control mechanisms and methods of breaking dormancy have been reviewed. Dormancy is an important survival mechanism that favors propagation and dissemination of a species but it has very strong negative consequences in crop production as it imposes heterogeneous and asynchronous seed germination. Many classification keys of dormancy exist, mainly because the process is generally controlled by more than one seed-dependent factor. For this reason induction and release of dormancy, remain a complex process in the life of plants. At the molecular level, induction and release of dormancy are regulated by a multifaceted network of transcriptional, translational, and epigenetic processes under an integrated control of environmental and hormonal signals. Though the process still needs elucidation, identification of the causes of dormancy is species-specific. In perspective, focus should be on diagnosing the dormancy types in each species with the objective to design a species specific protocol to suppress dormancy and improve on rapid and homogenous seed germination.

Author details

Tabi Kingsley Mbi[1*], Ntsomboh Godswill Ntsefong[2] and Tatah Eugene Lenzemo[1]

1 Department of Crop Production Technology, College of Technology, University of Bamenda, Cameroon

2 Faculty of Science, Department of Plant Biology, University of Yaounde I, Cameroon

*Address all correspondence to: kingotabimbi@yahoo.com

References

[1] Bewley JD. Seed germination and dormancy. The Plant Cell. 1997;**2**:1055-1066. DOI: 10.1105/tpc.9.7.1055

[2] Hilhorst HWM. A critical update on seed dormancy I, primary dormancy. Seed Science Research. 1995;**5**:61-73

[3] Donohue K, Rubio de Casas R, Burghardt L, Kovach K, Willis CG. Germination, post germination adaptation, and species ecological ranges. Annual Review of Ecology, Evolution, and Systematics. 2010;**41**:293-319. DOI: 10.1146/annurev-ecolsys-102209-144715

[4] Hilhorst HW. Standardizing seed dormancy research. Methods in Molecular Biology. 2011;**773**:43-52

[5] Bewley JD, Black M. Seeds Physiology of Development and Germination. 3rd ed. New York: Plenum Press; 1994. p. 445

[6] Baskin JM, Nan X, Baskin CC. A comparative study of seed dormancy and germination in an annual and a perennial species of Senna (Fabaceae). Seed Science Research. 1998;**8**:501-512. DOI: 10.1017/S0960258500004475

[7] Yarnia M, Tabrizi EFM. Effect of seed priming with different concentrations of GA_3, IAA and kinetin on Azarshahr onion germination and seedling growth. Journal of Basic Applied Science Research. 2012;**2**(3):2657-2661

[8] Foley ME, Fennimore SA. Genetic basis for seed dormancy. Seed Science Research. 1998;**8**:173-182. DOI: 10.1017/S0960258500004086

[9] Gubler F, Millar AA, Jacobsen JV. Dormancy release, ABA and pre-harvest sprouting. Current Opinion in Plant Biology. 2005;**8**:183-187

[10] Gordon AG. In: Heydecker W, editor. Seed Ecology. London: Butterworths; 1973. pp. 391-409

[11] Nikolaeva MG. Factors controlling the seed dormancy pattern. In: Khan AA, editor. The Physiology and Biochemistry of Seed Dormancy and Germination. Amsterdam, The Netherlands: North-Holland; 1977. pp. 51-74

[12] Foley ME, Black M, Bradford KJ, Vazquez RJ. Genetic model for dormancy in wild oat. In: Seed Biology: Advances and Applications. Wallingford, Oxon, UK: CABI Pub; 2000. pp. 323-327

[13] Gutterman Y. Seed Germination in Desert Plants. Adaptations of Desert Organisms. Berlin: Springer-Verlag; 1993. pp. 253

[14] Murugesan P, Mathur RK, Bijimol G, Kumar MR. Effect of extended dry heat treatment on germination and seedling growth in oil palm (*Elaeis guineensis*, Jacq) var. dura mother palms. Journal of Plantation Crops. 2008;**36**(1):45-48

[15] Liengsiri C, Hellum AK. Effects of temperature on seed *germination* in *Pterocarpus macrocarpus*. Journal of Seed Technology. 1988;**12**:66-75

[16] Rees AR. High-temperature pre-treatment and germination of seed of oil palm, (Elaeis guineensis Jacq.). Annals of Botany. 1962;**26**:569-581. DOI: /10.1093/oxfordjournals.aob.a083816

[17] Baskin CC, Zackrisson O, Baskin JM. Role of warm stratification in promoting germination of seeds of Empetrum hermaphroditum (Empertraceae), a circumboreal species with a stony endocarp. American Journal of Botany. 2000;**89**:486-493

[18] Holmquist JD, Popenoe J. The effect of scarification on the germination of seed of *Acrocomia crispa* and *Arenga engleri*. Principes. 1967;**11**:23-25

[19] Mayer AM, Poljakoff-Mayber A. The Germination of Seeds. 4th ed. Oxford: Pergamon Press; 1989

[20] Obroucheva NV, Antipova OV. Physiology of the initiation of seed germination. Russian Journal of Plant Physiology. 1997;**44**:250-264

[21] Bradbeer JW. Seed Dormancy and Germination. Glasgow: Blackie A and P; 1988. p. 146

[22] Hussey G. Analyses of the factors controlling the germination of seeds of oil palm (*Elaeis guineensis* Jacq.). Bottomline. 1958;**22**:259-284

[23] Davies RI, Pritchard HW. Seed storage and germination of the palms Hyphaenethebaica, *H. petersiana* and Medemia argun. Seed Science and Technology. 1998;**26**:823-828

[24] Moussa H, Margolis HA, Dube PA, Odongo J. Factors affecting the germination of doum palm (*Hyphaene thebaica* Mart.) seeds from the semi-arid zone of Niger, West Africa. Forest Ecology and Management. 1998;**104**:27-41

[25] Kubitzki K, Ziburski A. Seed dispersal in flood plain forests of Amazonia. Biotropica. 1994;**26**:30-43. DOI: 10.2307/2389108

[26] Geneve RL. Impact of temperature on seed dormancy. HortScience. 2003;**38**:336-341. DOI: 10.21273/HORTSCI.38.3.336

[27] Toole VK. *Effects of light, temperature* and their interactions on the germination of seeds. Seed Science and Technology. 1973;**21**:339-396

[28] Denslow JS. Tropical rainforest gaps and tree species diversity. Annual Review of Ecology and Systematics. 1987;**18**:431-451. DOI: 10.1146/annurev.es.18.110187.002243

[29] Brown KE. Ecological studies of the cabbage palm, Sabal palmetto. III. Seed germination and seedling establishment. Principes. 1976;**20**:98-115

[30] Orozco-Segovia A, Batis AI, Rojas-Aréchiga M, Mendoza A. Seed biology of palms: A review. Palms. 2003;**47**:79-94

[31] Goldberg RB, Paiva G, Yadegari R. Plant embryogenesis: Zygote to seed. Science. 1994;**2666**:605-614. DOI: 10.1126/science.266.5185.605

[32] Cendán C, Sampedro L, Zas R. The maternal environment determines the timing of germination in *Pinus pinaster*. Environmental and Experimental Botany. 2013;**94**:66-72. DOI: 10.1016/j.envexpbot.2011.11.022

[33] Figueroa R, Herms DA, Cardina J, Doohan D. Maternal environment effects on common groundsel (*Senecio vulgaris*) seed dormancy. Weed Science. 2010;**58**:160-166. DOI: 10.1614/WS-D-09-00006.1

[34] Castro J. Short delay in timing of emergence determines establishment success in *Pinus sylvestris* across microhabitats. Annals of Botany. 2006;**98**:1233-1240. DOI: 10.1093/aob/mcl208

[35] Hendrix SD. Variation in seed weight and its effects on germination in *Pastinaca sativa* L. (Umbelliferae). American Journal of Botany. 1984;**71**:795-802. DOI: 10.1002/j.1537-2197.1984.tb14144.x

[36] Kucerna B, Cohn MA, Leubner-Metzger G. Plant hormone

interactions during seed dormancy release and germination. Seed Science Research. 2005;**15**:281-307. DOI: 10.1079/SSR2005218

[37] Cadman CSC, Toorop PE, Hilhorst HWM, Finch-Savage WE. Gene expression profiles of Arabidopsis cvi seed during cycling through dormant and non-dormant states indicate a common underlying dormancy control mechanism. The Plant Journal. 2006;**46**:805-822. DOI: 10.1111/j.1365-313X.2006.02738.x

[38] Finkelstein RR, Gampala SSL, Rock CD. Abscisic acid Signaling in seeds and seedlings. Plant Cell. 2002;**14**(Suppl):s15-s45. DOI: 10.1105/tpc.010441

[39] Koornneef M, Karssen CM. Seed dormancy and germination. In: Meyerowitz EM, Somerville CR, editors. Arabidopsis. Cold Spring Harbor, NY: Cold Spring Harbor Laboratory Press; 1994. pp. 313-334

[40] McCarty DR. Genetic control and integration of maturation and germination pathways in seed development. Plant Molecular Biology. 1995;**46**:71-93

[41] Schopfer P, Plachy C. Control of seed germination by abscisic acid III. Effect on embryo growth potential (minimum turgor pressure) and growth coefficient (Cell Wall extensibility) in *Brassica napus* L. Plant Physiology. Mar 1985;**77**(3):676-686. DOI: 10.1104/pp.77.3.676

[42] Leida C, Conejero A, Arbona V, Gomez-Cadenas A, Llacer G, Badenes ML, et al. Chilling-dependent release of seed and bud dormancy in peach associates to common changes in gene expression. PLoS One. 2012;7(5):e35777. DOI: 10.1371/journal.pone.0035777

[43] Muller K, Tintelnot S, Leubner-Metzger G. Endosperm-limited Brassicaceae seed germination: Abscisic acid inhibits embryo-induced endosperm weakening of *Lepidium sativum* and endosperm rupture of cress and Arabidopsis thaliana. Plant & Cell Physiology. 2006;**47**:864-877

[44] Singh DP, Jermakow AM, Swain SM. Gibberellins are required for seed development and pollen tube growth in Arabidopsis. The Plant Cell. 2002;**14**(12):3133-3147. DOI: 10.1105/tpc.003046

[45] Hedden P, Kamiya Y. Gibberellin biosynthesis: Enzymes, genes and their regulatio. Annual Review of Plant Physiology and Plant Molecular Biology. 1997;**48**:431-460. DOI: 1040-2519/97/0601-0431

[46] Hays DB, mandel R. Hormones in zygotic and microspore embryos of Brassica napus. Plant Growth Regulation. 2001;**35**(1):47-58. DOI: 10.1023/A:1013831116996

[47] Khan MM, Ungar IA. Alleviation of seed dormancy in the desert forb *Zugophyllum simplex* L. from Pakistan. Annals of Botany. 1997;**80**:395-400

[48] Çetinbaş M, Koyuncu F. Improving germination of *Prunus avium* L. seeds by gibberellic acid, potassium nitrate and thiourea. Horticultural Science. 2006;**33**(3):119-123

[49] White CN, Rivin CJ. Gibberellins and Seed Development in Maize. II. Gibberellin synthesis inhibition enhances abscisic acid Signaling in cultured embryos. Plant Physiology. 2000;**122**(4):1089-1098. DOI: 10.1104/pp.122.4.1089

[50] Yücel E, Yilmaz G. Effects of different alkaline metal salts (NaCl,

KNO_3), acid concentrations (H_2SO_4) and growth regulator (GA3) on the germination of *Salvia cyanescens*. Gujarat Science. 2009;**3**:123-127

[51] Gehan GM, Mona FAA. Effect of gibberellic acid and indole 3-acetic acid on improving growth and accumulation of phytochemical composition in Balanitesa egyptiaca plants. American Journal of Plant Physiology. 2011;**6**(1):36-43. DOI: 10.3923/ajpp.2011.36.43

[52] Yamaguchi S. Gibberellin metabolism and its regulation. Annual Review of Plant Biology. 2008;**59**:225-251. DOI: 10.1146/annurev.arplant.59.032607.092804

[53] Nambara E, Marion-Poll A. ABA action and interactions in seeds. Trends in Plant Science. 2003;**8**:213-217. DOI: 10.1016/S1360-1385(03)00060-8

[54] Bentsink L, Koornneef M. Seed dormancy and germination. In: Arabidopsis Book. 2008. p. 6e0119

[55] Vleeshouwers LM, Bouwmeester HJ, Karssen CM. Redefining seed dormancy: An attempt to integrate physiology and ecology. Journal of Ecology. 1995;**3**(6):1031-1037

[56] Eira MTS, Caldas LS. Seed dormancy and germination as concurrent processes. Revista Brasileria Fisiologia Vegetal. 2000;**12**:85-104

[57] Wareing PF. Germination and dormancy. In: Wilkins MF, editor. The Physiology of Plant Growth and Development. New York: McGraw-Hill; 1969. pp. 605-644

[58] Hilhorst HWM, Karssen CM. Dual effect of light on the gibberellin- and nitrate-stimulated seed germination of *Sisymbrium officinale* and *Arabidopsis thaliana*. Plant Physiology. 1988;**86**:591-597

[59] Khan AA. Primary, preventive and permissive roles of hormones in plant systems. The Botanical Review. 1975;**41**:391-420

[60] Carrillo-barral N, Rodriguez-Gacio M, Matilla AJ. Delay of germination-1 (DOG1): A key to understanding seed dormancy. Plants. 2020;**9**:480

[61] Li X, Chen T, Li Y, Wang Z, Cao H, Chen F, et al. ETR1/RDO3 regulates seed dormancy by relieving the inhibitory effect of the ERF12-TPL complex on DELAY OF GERMINATION1 expression. Plant Cell. 2019;**31**:832-847. DOI: 10.1105/tpc.1800449

[62] Bryant FM, Hughes D, Hassani-Park K, Eastmond PJ. Basic leucine zipper transcription factor 67 transactivates delay of germination 1 to establish primary seed dormancy in Arabidopsis. Plant Cell. 2019;**31**: 1276-1288. DOI: 10.1105/tpc.18.00892

[63] Baskin CC, Thompson K, Baskin JM. Mistakes in germination ecology and how to avoid them. Seed Science Research. 2006;**16**:165-168. DOI: 10.1079/SSR2006247

[64] Baskin CC, Baskin JM. A classification system for seed dormancy. Seed Science Research. 2004;**14**:1-16. DOI: 10.1079/SSR2003150

[65] Holdsworth MJ, Bentsink L, Soppe WJJ. Molecular networks regulating Arabidopsis seed maturation, after-ripening, dormancy and germination. New Phytologist. 2008;**179**:33-54

[66] Bolingue W, Ly VB, Leprince O, Buitink J. Characterization of dormancy

behaviour in seeds of the model legume Medicago truncatula. Seed Science Research. 2010;**20**:97-107. DOI: 10.1017/S0960258510000061

[67] Andersson L, Milberg P. Variation in seed dormancy among mother plants, populations and years of seed collection. Seed Science Research. 1998;**8**:29-38. DOI: 10.1017/S0960258500003883

[68] Karssen CM, Brinkhorst-van der Swan DLC, Breekland AE. Induction of dormancy during seed development by endogenous abscisic acid: Studies on abscisic acid deficient genotypes of *Arabidopsis thaliana* (L.). Planta. 1983;**157**:158-165

[69] Sébastien B. An integrated overview of seed development in *Arabidopsis thaliana* ecotype WS. Plant Physiology and Biochemistry. 2002;**40**(2):151-160. DOI: 10.1016/S0981-9428(01)01350-X

[70] Finch-Savage WE, Leubner-Metzger G. Seed dormancy and the control of germination. New Phytologist. 2006;**171**:501-523. DOI: 10.1111/j. 1469-8137.2006.01787.x

[71] Baskin CC, Baskin JM. Seed dormancy in wild flowers. In: Flower Seeds: Biology and Technology. Wallingford: CAB International; 2005. pp. 163-185

[72] Baskin CC, Baskin JM. Breaking seed dormancy during dry storage: A useful tool or major problem for successful restoration via direct seeding? Plants. 2020;**9**(5):636. DOI: 10.3390/plants9050636

[73] Wannan BS, Quinn CJ. Pericarp structure and generic affinities in Anacardiaceae. Botanical Journal of the Linnean Society. 1990;**102**:225-252. DOI: 10.1111/j.1095-8339.1990.tb01878.x

[74] Baskin CC. Breaking physical dormancy in seeds: Focussing on the lens. New Phytologist. 2003;**158**:229-232. DOI: 10.1046/j.1469-8137.2003.00751.x

[75] Kildisheva Olga A, Anthony SD, Bassam AH. A hard seed to crack evaluating dormancy-breaking techniques for mamane. Native Plants Journal. 2013;**14**(3):243-248. DOI: 10.3368/npj.14.3.243

[76] Baskin CC, Per Milberg LA, Baskin JM. Non-deep simple morphophysiological dormancy in seeds of the weedy facultative winter annual Papaver rhoeas. Weed Research. 2002;**42**(3):194-202. DOI: doi/abs/10.1046/j.1365-3180.2002.00275.x

[77] Yang QH, Ye WH, Yin XJ. Dormancy and germination of Areca triandra seeds. Sci Hortic-Amsterdam. 2007;**113**:107-111. DOI: 10.1016/j.scienta.2007.01.028

[78] Tabi-Mbi K, Ngando-Ebongue GF, Ntsomboh-Ntsefong G, Youmbi E. Effect of dry heat treatment along with some dormancy breaking chemicals on oil palm seed germination. South African Journal of Botany. 2017;**112**:489-493. DOI: 10.1016/j.sajb.2017.06.023

[79] Khasa PD. Scarification of limba seeds with hot water, bleach and acid. Tree Planters Note. 1992;**43**(4):150-152

[80] Das N. The effect of seed sources variation and Presowing treatments on the seed germination of *Acacia catechu* and *Elaeocarpus floribundus* species in Bangladesh. International Journal of Forestry Research. 2014;**1**:1-8. DOI: 10.1155/2014/984194

[81] Ameer JBM, Vijayakumar A, Selvaraju P. Standardization of seed dormancy breaking treatment in Senna (*Cassia auriculata*). Journal

of Plant Breeding and Crop Science. 2013;**5**(11):220-223

[82] Meerow AW. Palm seed germination. Co-operative Extension Service Bulletin. 1991;**274**:1-17. DOI: 10.4236/as.2016.73016

[83] Bonner FT, McLemore BF, Barnett JP. Presowing treatment of seed to speed germination. In: Bonner FT, editor. Seeds of Woody Plants in the United States, Agriculture Handbook. Washington D.C: USDA; 1974

[84] Hanson AD. The effect of imbibition drying treatment on wheat seeds. The New Phytologist. 1973;**72**:1063-1073. DOI: 10.1111/j.1469-8137.1973.tb02083.x

[85] Barnett JP, McLemore BF. Improving storage of spruce pine seed. Tree Planters' Notes. 1967;**18**(2):16

[86] Berrie AMM, Drennan DSH. The effects of hydration-dehydration on seed germination. The New Phytologist. 1971;**70**:135-142. DOI: 10.1111/j.1469-8137.1971.tb02518.x

[87] Tabi-Mbi K, Ntsomboh-Ntsefong G, Tonfack LB, Ngando-Ebongue GF, Youmbi E. Optimization of rhizogenesis in oil palm vitro plantlets derived from direct organogenesis of mature zygotic embryos (MZE). Advanced Plants and Agricultural Research. 2018;**8**(3):223-235. DOI: 10.15406/apar.2018.08.00319

[88] Hu C, Wang P. Embryo culture: Technique and applications. In: Evans DE, Sharp WR, Ammirato PV, editors. Handbook of Plant Cell Culture. Vol. 4. New York: Macmillan; 1986. pp. 43-96

[89] Tamaki M, Urasaki N, Nakamura I, Motomura K, Adaniya S. Shortening the breeding cycle of papaya (*Carica papaya* L.) by culturing embryos treated with ethrel. Plant Cell, Tissue and Organ Culture (PCTOC). 2011;**106**(2):225-233. DOI: 10.1007/s11240-010-9910-6

Chapter 4

Factors That Cause Seed Dormancy

Mohammed Shu'aibu Abubakar
and Muhammad Lawal Attanda

Abstract

Seed dormancy is a state in which seeds do not germinate despite the presence of all of the necessary conditions (temperature, humidity, oxygen, and light). It is caused by hard seed coat impermeability or a lack of supply and activity of the enzymes required for germination. The dormancy of seeds presents a practical problem of considerable economic importance. Plant growers are often interested in securing seed that will germinate soon after it is harvested. To overcome dormancy, organic material is subjected to a variety of physical and chemical pretreatments. Some plant species have both physical and internal dormancy, making it difficult to produce high-frequency healthy seedling growth, despite the fact that seed sprouting and the generation of healthy seedlings is a requirement for plant output. The Chapter is intended to present the basic information on the seed dormancy which would be of relevance to the seed growers and scientists during seed handling process. Seed dormancy is of great concern to scientists therefore it is a research area of interest. All the viable seeds have capacity to germinate if placed under suitable conditions necessary for germination. But some seeds fail to germinate for sometimes even if placed under the condition favorable for germination.

Keywords: seed coat, seed dormancy, seed germination, types of seed dormancy, breaking seed dormancy, factors responsible for seed dormancy

1. Introduction

The history of seed is the evolution of crop cultivation for human sustenance and survival. Over time this progression had been achieved through introduction, development and release of varietal of seeds using various well-known techniques of selection, hybridization and polyploidisation. However, all these stages of seed development have little value if main aim of seed development is not achieved due to a block to the completion of germination widely known as seed dormancy. Seed dormancy is an inactive status of seed and its property that normally describes the environmental factors in which the seed would germinate. This phenomenal is determined by both seed genetics, hormones, and substantial environmental influence in seed maturation environment. However, seed dormancy could be considered simply as a block to the completion of germination of an intact viable seed under

IntechOpen

favorable conditions [1–3]. All climatic regions have cases of seed dormancy but with varying divergent responses in adaptation to avoid hostile climate conditions for germination. The adaptation is timed to avoid unfavorable weather for germination and subsequent plant establishment and reproductive growth. All viable seeds have the ability to sprout when placed in the proper germination circumstances. However, some seeds, even when placed under ideal conditions for germination, fail to germinate. This could be due to internal reasons or a necessity for specific external conditions. During this time, the seeds' growth is halted, and they are said to be in a rest or inactive state, and this occurrence is known as seed dormancy [4]. Seed dormancy is among the least implicit phenomena in the field of seed biology particularly with the issues of clear description, unambiguous definitions and distinct types and classifications of the seed dormancy as well as the vital distinction of the different methods to terminate dormancy or induce germination and factors responsible for the mechanisms. In this chapter an attempt is made to discourse these issues in perspective of types and methods of breaking seed dormancy and factors responsible for dormancy.

2. Meaning of seed dormancy

2.1 Definitions of seed dormancy

Seed dormancy is an internal condition of a viable seed which constraint its germination despite enable growing conditions of suitable temperature and moisture availability [4]. Therefore, seed dormancy can be simply described as a resting state of a viable seed that must be broken either by time or deliberate conditions before the seed germinates at temperature and moisture levels suitable for required growth. In operational term, dormancy is a block to the completion of germination of an intact viable seed with suitable growth conditions. The block evolves based on the species of the seeds and climatic conditions in the prevailing environment [5–8]. Also, seed dormancy is a state in which seeds do not germinate despite the presence of all of the necessary conditions (temperature, humidity, oxygen, and light). It is caused by hard seed coat impermeability or a lack of supply and activity of the enzymes required for germination. Dormancy is a significant limiting factor in the production of many field crops. To overcome dormancy, organic material is subjected to a variety of physical and chemical pretreatments. Some plant species have both physical and internal dormancy, making it difficult to produce high-frequency healthy seedling growth, despite the fact that seed sprouting and the generation of healthy seedlings is a requirement for plant output. However, the constraint in definitions of the seed dormancy is inability to observe it in no other measure than the absence of germination [9, 10]. This is because a state of dormancy can take a value between maximum dormancy and non-dormancy. Therefore, dormancy is not typically associated with the absence of germination in seed rather it is a characteristic of the seed that determines the conditions essential for germination.

2.2 Seed dormancy and germination

A dormant seed has no capacity to germinate even in the presence of favorable environmental factors and habitable conditions due to either non-viable, empty of embryo or dormancy [4, 5]. Many types of seeds, even when they appear to be ripe, fail to germinate even when all environmental parameters are favorable. The embryo's resumption of growth in such seeds is halted by the circumstances within the seeds. The state of slowed growth of seeds or other plant organs caused by internal factors is known as dormancy, but it is also known as the rest period. Seed dormancy is of considerable advantage to the plant. It enables the embryo to safely pass the unfavorable part of the year and germinate when the conditions are suitable for the establishment of the seedlings [3, 5]. In nature, dormancy period coincides with the unfavorable period for the seedling of the species. In other words, a completely non-dormant seed has the capacity to germinate over the widest range of normal physical environmental factors possible for the genotype. Therefore, a suspected case of seed dormancy arises when fresh un-germinated seeds remain at the end of a germination trial period that required some particular conditions of either endogenous, exogenous or combination of both factors to be satisfied. For seed to germination there are needs for basic requirements of water, oxygen and appropriate temperature, in consideration of other factors such as light and hormones. However, seed germinations are generally in three phases, starting with the uptake of water by imbibition of the dry seed which leads to embryo expansion [11–14].

First phase:

This phase is known as Imbibition where there is rapid initial water uptake by the hard seed in the process of hydration of protoplasm which would be followed by activation and synthesis of enzymes.

Second phase:

This is plateau phase where increase in water uptake leads to increase in respiration of embryo preceding activation and synthesis of enzymes. This increase synthesis of nucleic acids and proteins with release of hormones from the embryo.

Third phase.

This is embryo elongation phase. It's a point where radicle protrusion which is generally accepted to be sufficient for completion of germination is visible. This stage entails increased cell enlargement, cell division hydrolysis of reserve food and utilization of soluble organic substance by developing embryo for the growth of the radicle into root and plumule into shoot.

3. Types of seed dormancy

There are several categories of seed dormancy classification in literatures. However, seed dormancy may be classified based on the following schemes [15];

i. Mode of seed origin

ii. Mode of seed structure

iii. Mode of action

- *Mode of seed origin:* Based on the seed provenance dormancy can be classified into two main groups;

    a. **Primary dormancy:** this is an inherent dormancy status of a maturing seed when its fully developed on the parent plant. This is when the seed is released from the mother plant in a dormant state.

b. **Secondary dormancy**: This is when the seed released from a parent plant becomes dormant as a result of environmental conditions. This is regarded as induced dormancy whose germination of the seeds have been inhibited. Secondary dormancy develops in some non-dormant and post-dormant seeds exposed to unfavorable germination conditions, such as high temperatures. It's brought on by circumstances that arise after the seed has been disseminated. The mechanisms of secondary dormancy are unknown; however, they may entail the reduction of sensitivity in plasma membrane receptors. Many species of plants release their seeds late in the year when the soil temperature is too low for germination or when the environment is too dry for germination. These seeds will germinate if they are gathered and sown in an environment that is warm and/or moist enough. Non-dormant seeds released late in the growing season in natural settings season, wait until the soil temperature rises in the spring, or in the event of seeds scattered during dry seasons, wait until it rains and the soil is sufficiently moist [16–18].

- *Mode of seed structure:* Seed dormancy can be categorized into two main groups based on the parts of the seed where the dormancy affects and control.

a. **Embryo induced dormancy:** the dormancy is within the embryo itself and usually caused by undeveloped or complete dormant embryo

b. **Coat induced dormancy:** This is the dormancy which resides within the structure (seed coat) enclosing the embryo and caused by extreme hardness of the seed coat.

- *Mode of action:* This mode of classification is based on the action of dormancy that is shown to be determined by both the physiological and morphological properties of the seed.

Generally, seed dormancy could either be exogenous or endogenous dormancy.

3.1 Exogenous dormancy

Exogenous dormancy is caused by conditions outside the embryo and is often broken down into three subgroups [19–22]:

i. **Physical dormancy;** this is caused by impermeability of layers of macrosclereld cells and mucilaginous outer cells to water. The movement of water is restrained by hardened endocarp of the seeds. This happens when seeds are impervious to water or gas exchange. Legumes are a good example of physically inactive seeds since they have a low moisture content and the seed coat prevents them from absorbing water. Water can be absorbed by chipping or splitting the seed coat or other coatings. Impermeability is frequently generated by an exterior cell layer made up of macrosclereid cells or a mucilaginous cell layer. A stiffened endocarp is the third source of seed coat impermeability. During the latter stages of seed development, seed coverings that are impervious to water and gases form.

ii. **Mechanical dormancy;** when seed coats or other coverings are too tough to allow the embryo to expand during germination, this happens. Previously, this method of dormancy was ascribed to a number of species, however endogenous components were discovered to be responsible for their hibernation. Physiological dormancy caused by poor embryo development potential is one of these endogenous facts.

iii. **Chemical dormancy**; this includes growth regulators found in the embryo's surrounding tissues. They can be washed or soaked out of the seed's tissues, or they can be deactivated in various ways. Rainwater or snowmelt removes other compounds that hinder seeds from germinating.

3.2 Endogenous dormancy

Endogenous dormancy is divided into physiological dormancy, morphological dormancy, and combined (morpho-physiological) dormancy, and is typically triggered by factors within the embryo itself [23–25].

i. **Physiological dormancy;** this is the most common type of seed dormancy in seed biology. An embryo growth is retarded due to inhibitors changes and seed germination prevented. It is as a result of seed not satisfying certain physiological conditions necessary for germination. Physiological dormancy stops embryos from growing and seeds from germinating until chemical changes take place. Inhibitors are among these compounds, and they often slow embryo growth to the point where it is unable to break through the seed coat or other tissues. An increase in germination rate following the application of gibberellic acid (GA3) or after Dry after-ripening or dry storage indicates physiological dormancy. It's also useful when latent seed embryos are removed and healthy seedlings result; or when up to 3 months of cold (0–10°C) or warm (>15°C) stratification improves germination; or when dry after-ripening reduces the time spent cold stratifying. Scarification increases germination in some seeds, indicating physiological dormancy.

ii. **Morphological dormancy;** this manifests in seeds with embryos that are under developed in size but clearly with differentiated cotyledons, hypocotyls and radical axis. The embryos in morphological dormancy are normally under favorable conditions only require time to grow and germinate because they are not physiologically dormant. Underdeveloped or undifferentiated embryo. Some seeds have completely developed embryos that need to expand more before seed germination, whereas others have embryos that have not yet differentiated into distinct tissues when the fruit ripens. Immature embryos - certain plants release their seeds before the embryos' tissues have fully differentiated, and the seeds ripen after they take in water while on the ground; germination might take weeks or months.

iii. **Combined (Morpho-physiological) dormancy;** Combined dormancy occurs in some seeds, where dormancy is caused by both exogenous (physical) and endogenous (physiological) conditions. This is a dormancy where the seeds have under development in size but distinctly differentiated embryos in addition to components of physiological dormancy. The seeds under this dormancy require time to enable the embryo to grow and germinate as well as dormancy breaking treatment. Seeds are dormant both morphologically and physiologically. Morpho-physiological dormancy, also known as morphophysiological dormancy, occurs when seeds with immature embryos exhibit physiological dormancy. As a result, dormancy-breaking procedures and a period of time to generate fully formed embryos are required for these seeds.

a. Intermediate simple

b. Deep simple

c. Deep simple epicotyl

d. Deep simple double

e. Intermediate complex

f. Deep complex

4. Methods of breaking seed dormancy

Various methods have been used by seed scientist and technologists to break the dormancy of seed. Simple and widely used methods are [26]:

4.1 Scarification

Scarification; is a term used to describe any physical or chemical treatment that weakens the seed coat. When a hard seen coat imposes dormancy, such as in legumes like Cajanus cajan (tur), gram, and others, the scarification method is used. There are several ways to shatter the hard seed coat with this procedure, including [26–28]:

1. The seeds are manually rubbed on sand paper. When rubbing seeds such as green gram and subabool, care should be taken not to injure the axis of the seed.

2. When the seed coat is too hard, especially if it is of a woody type, the seed coat must be broken off completely. Rubber (Havea spp.) seed, for example, is used to make India teak wood.

3. Soaking treatment: Eliminates seed coat impermeability by soaking hard seed coats in a concentrated or diluted sulfuric acid solution for 1 to 60 minutes. Cotton seeds, for example, and India teak wood seeds.

4.2 Temperature treatments

Temperature Treatments; when dormancy is caused by embryo factor, the seed is incubated at a low temperature (0–5°C) on a substratum for 3 to 10 days, allowing it to reach its optimum temperature. Germination necessitates the presence of this ingredient. - Take mustard, for example (Brassica campestrits). Before germinating at the proper temperature, some seeds required a brief period of incubation (from a few hours to one to five days) at 40 to 50°C. (When using this approach, make sure the seed has a moisture level of no more than 15%, for example, paddy) (Oryza sativa). Breaking hard-seed ness in legumes with hot water treatment is also an effective strategy. The seeds are steeped in water at a temperature of 80°C for 1–5 minutes in this approach before putting for germination (depending up on the type of seed) [29].

4.3 Light treatments

Light Treatments; some seeds do not germinate in dark thus are providing with continuous or periodic exposure of light is essential for example Lettuce (*Lactuca sativa*) required red light (660 nm) or white light is essential for germination to occur [20].

4.4 Treatments with growth regulators and other chemicals

Treatments with growth regulators and other Chemicals; the presence of germination inhibitors may cause endogenous dormancy. Low-dose growth regulators (Gibberellins, Cytokinins, and Ethylene) may be used to break seed dormancy. Gibberellins and kinetics are the most extensively utilized growth regulators; for example, presoaking seed treatment with GA3 at a concentration of 100 ppm has been employed to break seed dormancy in sorghum seeds. Potassium nitrate (0.2 percent) and thio – urea (0.5 to 3 percent) is commonly used to break seed dormancy in oat (Avena sativa), barley (Hordeum vulgare), and tomato (Solanum lycopersicum) (Solanum Lycopersicum) [30].

Several methods have been devised for breaking the dormancy of seeds and for shortening the period of dormancy so that they may germinate promptly. Whenever dormancy results of any of the causes inherent in the seed coats it can be interrupted by scarification. For example, machine thrashed legumes seeds usually show a higher percentage of germination than those that have been harvested by hand. Strong mineral acids have been used successfully to interrupt seed dormancy caused by resistant or impermeable seed coats. Soaking the seeds in certain chemicals like potassium nitrate, ethylene, chlorohydrine, thiourea or in certain plant hormones is known to break dormancy. After ripening of many seeds occurs more rapidly when they are kept at low temperatures than when stored at higher temperatures. Temperatures from 5–10°C for two or three months are effective with seeds of conifers. Under natural conditions, seed dormancy is gradually overcome by processes such as weakening of the seed coat by the digestive juices in the alimentary canal of fruit-eating birds and other animals, or in still due to the action of microbes or due to mechanical abrasions. Dormancy of seeds is also broken by subjecting the seeds alternately to relatively low and high temperatures. Light is also considered as means of breaking dormancy of seeds. Seeds of sweet clover (Melilotus alba) and alfalfa (Medicago sativa) showed greatly improved germination after being subjected to hydraulic pressures of 2000 atmphere at 18°C. The following are some general approaches for breaking seed dormancy [31–33]:

1. **Dry storage:** It is often sufficient to store the sample in a dry spot for a short amount of time for species where dormancy is naturally short.

2. **Pre-chilling:** The germination duplicates are placed in contact with the moist substratum and held at a low temperature for a length of time before being raised to the appropriate temperature.

3. **Pre-heating:** The germination replicates should be heated for up to seven days at a temperature of not more than 40°C with free air circulation before being placed under the appropriate germination conditions. It may be essential to lengthen the pre-heat duration in some circumstances.

4. **Light:** The seed should be illuminated during at least 8 hours in every 24 hours cycle and during the high temperature period when the seeds are germinated at alternating temperatures. The light intensity should be approximately 750–1250 lux from cool white lamps.

5. **Potassium nitrate (KNO_3):** The germination substratum may be moistened with a 0.2 per cent solution of KNO_3, prepared by dissolving 2 gm KNO_3 in one liter of water. The substratum is saturated at the beginning of the test, but water is used for moistening it thereafter.

6. **Gibberellic Acid (GA_3):** The germination substratum may be moistened with 500 ppm solution of GA_3, prepared by dissolving 500 mg GA_3, in one liter of water. When dormancy is weak, 200 ppm may be sufficient. When the problem is severe, a 1000 ppm solution may be utilized. When the concentration is greater than 800 ppm, a buffer of 0.01 M in distilled water can be employed.

7. **Pre-washing:** When germination is affected by a naturally occurring substance in the seeds, which acts as an inhibitor it may be removed by washing the seeds in running water at room temperature (25°C) before the germination test is made. After washing the seeds should be dried back at room temperature (25°C).

8. **Removal of structures around the seed:** Germination can be promoted by removing outer structures such as involucre of bristles or lemma.

9. **Disinfection of the seed:** A fungicide treatment may be applied before planting the seed for germination, when the seed is known not to have received such a treatment.

10. **Soaking:** Seeds with hard seed coats may germinate more readily after soaking for 24–48 hours in water. The seed should be planted for germination immediately after soaking.

11. **Mechanical scarification:** Breaking the dormant condition may be as simple as piercing, chipping, filing, or sand papering the seed coat. Scarifying the seed coat at a proper location is necessary to avoid harming the embryo and the resultant seedling. The region of the seed coat directly above the cotyledon tips is the optimum location for mechanical scarification.

12. **Acid scarification:** With some spices of seeds steeped in concentrated H_2SO_4 until the seed coat becomes pitted, digestion in concentrated H_2SO_4 is successful. Digestion may be rapid or take more than one hour but the seeds should be examined every few minutes. After digestion, seeds must be thoroughly washed in running water before the plant for germination.

5. Factors responsible for seed dormancy

5.1 Impermeability of seed-coats to water

The seed coats of many species are completely impermeable to water at the time of their maturity. This condition is very common in the seeds of many legumes, (example, clovers, alfalfa) of the water lotus, and of the morning glory. Germination fails to occur until water penetrates through the seed coats [28–32, 34].

5.2 Mechanically resistant seed coats

In some seeds as those of mustard (Brassica), pigweed (Amaranthus), shepherd's purse (Capsella), the seed-coats are so strong that they do not yield to the pressure of the expanding embryo. The embryos of these seeds have no dormant period and will grow readily if the seed coats are removed.

5.3 Seed-coats impermeable to oxygen

The two seeds in a cocklebur (Xanthium) fruit are not dormant in the same way. The lower seed normally germinates in the spring following maturity in natural settings, while the top seed remains dormant until the following year. The impermeability of the seed coverings to oxygen has been shown to be the cause of dormancy in these seeds.

5.4 Rudimentary embryos

In plants like ginkgo (Ginkgo biloba), European ash (Fraxinus), holly (Ilex) and many orchids, the embryo is unorganized when the seed is shed and attains full development before it germinates.

5.5 Dormant embryos

Even when the embryos are fully grown when the seed is ripe, many species' seeds fail to germinate, even when the environmental conditions are ideal. The physiological state of the embryo causes dormancy in such seeds. Even if the seed coverings are removed, the embryos of such seeds will not grow when they first mature. During the period of dormancy, some physiological changes called after-ripening occur in the embryo before the seed is capable of germination. The seeds of apple, peach, iris and pine belong to this group. In nature, after-ripening occurs in winter and the seeds formed in autumn germinate the coming spring.

5.6 Germination inhibitors

Many species' seeds fail to germinate even when the embryos are completely developed when the seed is ripe, even though the environmental conditions are excellent. In such seeds, dormancy is caused by the physiological state of the embryo. The embryos of such seeds will not grow when they initially mature, even if the seed covers are removed. Inhibitors may be present in the embryo (example, in Xanthium), endosperm, (example, in Iris) or in the seed coat (example, in Cucurbita). Abscissic acid (ABA) is one of the most commonly detected inhibitors of germination.

6. Conclusion

Seed dormancy is a state in which seeds do not germinate despite the presence of all of the necessary circumstances (temperature, humidity, oxygen, and light), and is caused by hard seed coat impermeability or a lack of supply and activity of the enzymes required for germination. The dormancy of seeds presents a practical problem of considerable economic importance. Plant growers are often interested in securing seed that will germinate soon after it is harvested. Dormancy is a significant limiting factor in the production of many field crops. To overcome dormancy, organic material is subjected to a variety of physical and chemical pretreatments. Some plant species have both physical and internal dormancy, making it difficult to produce high-frequency healthy seedling growth, despite the fact that seed sprouting and the generation of healthy seedlings is a requirement for plant output. The Chapter is intended to present the basic information on the seed dormancy which would be of relevance to the seed growers (farmers) and scientists during seed handling process. Actually, seed dormancy is of great concern to scientists therefore it is a research area of interest. All the viable seeds have capacity to germinate if placed under suitable conditions necessary for germination. But some seeds fail to germinate for sometimes even if placed under the condition favorable for germination.

Author details

Mohammed Shu'aibu Abubakar* and Muhammad Lawal Attanda
Department of Agricultural and Environmental Engineering, Bayero University, Kano, Nigeria

*Address all correspondence to: msabubakar.age@buk.edu.ng

References

[1] Webb DP, Wareing PF. Seed dormancy in acer: Endogenous germination inhibitors and dormancy in Acer pseudoplatanus L. Planta. 1972;**104**:115-125

[2] Williams PM, Ross JD, Bradbeer JW. Studies in seed dormancy: 4. The abscisic acid content of the seeds and fruits of Corylus avellana L. Planta. 1973;**110**:303-310

[3] Bonner FT. Glossary of Seed Germination Terms for Tree Seed Workers. Gen. Tech. Rep. SO-49. New Orleans: USDA Forest Service, Southern Forest Experiment Station; 1984. p. 4

[4] Soltani F, Kashi A, Arghavani M. Effect of magnetic field on Asparagus officinalis L. seed germination and seedling growth. Seed Science and Technology. 2006;**34**:349-353

[5] Sondheimer E, Tzou DS, Galson EC. Abscisic acid levels and seed dormancy. Plant Physiology. 1968;**43**:1443-1447

[6] Baskin CC, Baskin JM. Germination ecology of seeds with nondeep physiological dormancy. In: Baskin CC, Baskin JM, editors. Seeds: Ecology, Biography, and Evolution of Dormancy and Germination. San Diego, California: Academic Press; 1998. pp. 57-64

[7] Emery DE. Seed Propagation of Native California Plants. Santa Barbara, CA: Santa Barbara Botanic Garden; 1987

[8] Florez M, Carbonell MV, Martinez E. Exposure of maize seeds to stationary magnetic fields: Effects on germination and early growth. Environmental and Experimental Botany. 2007;**59**:68-75

[9] Enu-Kwesi L, Dumbroff EB. Changes in abscisic acid in the embryo and covering structures of Acer saccharum during stratification. Zeitschrift für Pflanzenphysiologie. 1978;**86**:371-377

[10] Tillberg E, Pinfield NJ. Changes in abscisic acid levels during, after-ripening and germination of Acer platanoides L. seeds. New Phytologist. 1982;**92**:167-172

[11] Dale JE. The control of leaf expansion. Annual Review of Plant Physiology and Plant Molecular Biology. 1988;**39**:267-295

[12] Singh Z, Browning G. The role of ABA in the control of apple seed dormancy re-appraised by combined gas chromatography-mass spectrometry. Journal of Experimental Botany. 1991;**42**:269-275

[13] Szczotka Z. Changes in the activity of indole acetic acid and abscisic acid in the embryo axes and cotyledons of Quercus borealis Michx. And Quercus robur L. acorns stored under controlled conditions. Arboretum Kornickie. 1977;**22**:257-273

[14] Ogawa K, Iwabuchi M. A mechanism for promoting the germination of Zinnia elegans by hydrogen peroxide. Plant and Cell Physiology. 2001;**42**:286-291

[15] Kanecko Y, Morohashi Y. The effect of sodium hypochlorite treatment on the development of α-amylase activity in mung bean cotyledons. Plant Science. 2003;**164**:287-292

[16] Murashige T, Skoog F. A revised medium for rapid growth and bioassays with tobacco tissue cultures. Physiologia Plantarum. 1962;**15**:431-497

[17] Racoppi F. Domestic Bleaches Containing Sodium Hypochlorite.

Procter and Gamble. Rome: Italia SPA Product Development Department; 1990

[18] Chaomei Z, Yanlin M. Irradiation induced changes in enzymes of wheat during seed germination and seedling growth. Acta Agriculturae Nucleatae Sinica. 1993;7:93-97

[19] Schull W. Temperature and rate of moisture intake in seeds. Botanical Gazette. 1920;**69**:361

[20] Hsiao AI, Hans JA. Application of sodium hypochlorite seed viability test to wild oat populations with different dormancy characteristics. Canadian Journal of Plant Science. 1981;**61**:115-122

[21] Hsiao AI, Quick AW. Action of sodium hypochlorite and hydrogene peroxide on seed dormancy and germination of wild oats, Avena fatua L. Weed Research. 1984;**24**:411-419

[22] Sunderland N. Cell division and expansion in the growth of the leaf. Journal of Experimental Botany. 1960;**11**:68-80

[23] Yildiz M, Özgen M. The effect of a submersion pretreatment on in vitro explant growth and shoot regeneration from hypocotyls of flax (Linum usitatissimum). Plant Cell, Tissue and Organ Culture. 2004;**77**:111-115

[24] Gunckel JE, Sparrow AH. In: Ruhland W, editor. Encycl. Plant Physiology, in Ionizing Radiation: Biochemical, Physiological and Morphological Aspects of their Effects on Plants. Berlin: Springer-Verlag; 1961. pp. 555-611

[25] Kim JH, Chung BY, Kim JS, Wi SG. Effects of in planta gamma-irradiation on growth, photosynthesis, and antioxidative capacity of red pepper (Capsicum annuum L.) plants. Journal of Plant Biology. 2005;**48**:47-56

[26] Wi SG, Chung BY, Kim JH, Baek MH, Yang DH, Lee JW, et al. Ultrastructural changes of cell organelles in Arabidopsis stem after gamma irradiation. Journal of Plant Biology. 2005;**48**:195-200

[27] Radhadevi DS, Nayar NK. Gamma rays induced fruit character variations in Nendran, a variety of banana (Musa paradasiaca L.). Geobios. 1996;**23**:88-93

[28] Kumari R, Singh Y. Effect of gamma rays and EMS on seed germination and plant survival of Pisum sativum L. and Lens culinaris medic. Neo Botanica. 1996;**4**:25-29

[29] Maherchandani N. Effects of gamma radiation the dormant seeds of Avena fatua. Radiation Botany. 1975;**15**:439-443

[30] Telci C, Yıldız M, Pelit S, Önol B, Erkılıç EG, Kendir H. The effect of surface-disinfection process on dormancy-breaking, seed germination, and seedling growth of Lathyrus chrysanthus Boiss. Under in vitro conditions. Propagation of Ornamental Plants. 2011;**11**:10-16

[31] Böhm J. Terrestrial orchid seed handling: Growing orchids of temperate climates from seed. In: Seed Pretreatment Protocol. Homnurg-Saar, Germany: Institute of Physiology, Medical Faculty, University of Saarland; 2003

[32] Kuzin AM, Vagabova ME, Vilenchik MM, Gogvadze VG. Stimulation of plant growth by exposure to low-level gamma radiation and magnetic field, and their possible mechanism of action. Environmental and Experimental Botany. 1986;**26**:163-167

[33] Charbaji T, Nabulsi I. Effect of low doses of gamma irradiation on in vitro

growth of grapevine. Plant Cell, Tissue and Organ Culture. 1999;**57**:129-132

[34] Melki M, Marouani A. Effects of gamma ray's irradiation on seed germination and growth of hard wheat. Environmental Chemistry Letters. 2010;**8**:307-310

Chapter 5

In Vitro Seed Germination and Seedling Development of Two Avocado Varieties

Essoh Aimé Cesaire Elekou, Irene Perea-Arango, Luis María Suarez-Rodriguez and Rodolfo López-Gómez

Abstract

Avocado (*Persea americana* Mill.) is a tree native to central and eastern Mexico. A basal angiosperm of the Lauraceae family, it produces an oil-rich fruit that is appreciated worldwide for its nutritional value. Mexico is the world's leading avocado producer. Production is based mainly on the use of rootstocks of *Persea americana* var. drymifolia, a "Mexican native". The agronomic characteristics of the rootstock are key to avocado production. This work reports a germination method to obtain seedlings *in vitro* from two avocado varieties, *P. americana* var. drymifolia and West Indian *P. americana* var. americana. With this system, germination success rates of 100% were obtained in a maximum of five days, with homogeneous seedling development. This system could provide rootstock that improves the characteristics of agronomic programs and the selection of genetic material for avocado production.

Keywords: seed, *in vitro*, avocado fruit, *persea americana* var. drymifolia, cotyledon, seedling

1. Introduction

The formation, dispersal, and germination of seeds are very important events in the life-cycle of gymnosperm and angiosperm plants [1]. As a member of the basal Angiosperms, the avocado tree, of the Lauraceae family, is considered by numerous botanists to be one of the most primitive dicotyledonous plants [2]. They are species of economic importance that are cultivated in tropical and subtropical areas worldwide [3]. Three botanical varieties (races) have been reported, known as Mexican, Guatemalan, and West Indian [4–6].

Latin America is the area with the highest production in the world, with Mexico being the largest producer [7]. The "Hass" cultivar has the highest consumer acceptance, production, and commercialization indices. This cultivar results from the hybridization of the Mexican and Guatemalan varieties [8], but cultivation requires rootstock from one of the botanical varieties, usually obtained from seeds [9, 10].

However, seedlings grown in nursery systems may require 40–60 days to germinate after sowing [11] and may lack uniformity due to the high genetic variation characteristic of seeds. Thus, various authors have reported the importance of shortening germination times and favoring this process using *in vitro* systems. Previous studies have reported 94–99% success rates in the germination of whole seeds and embryos barely covered by 1.5 cm^2 of cotyledons in *in vitro* systems, with germination emerging at 3–10 days [12]. Various culture media compositions have also been evaluated with *Persea americana* var. drymifolia and *Persea americana* cv. Hass, reported average germination emergences of 20–27 days [13]. The advantages of *in vitro* germination may provide a solution to cases of the total inhibition of germination, foster increased germination rates, reduce the time required, and improve the homogenization of seedlings that can serve as rootstocks [14]. In this work, we report data on the germination and development of *in vitro* seedlings obtained from *P. americana var. drymifolia* and *Persea americana* var. americana.

2. Materials and methods

2.1 Biological material

Avocado seeds were obtained from two varieties of fruit. *P. americana var. drymifolia* were obtained from crops in Tingambato and local markets in Morelia, Michoacán, Mexico. *P. americana var. americana* seeds were obtained from local markets in Merida, Yucatán, Mexico. All fruits were in the ripe stage (**Figure 1**).

2.2 Seed disinfection treatment

The surface of the avocados was washed with soap and water, then the fruit was cut to obtain the seeds, which were rinsed with sterile purified water to eliminate remaining parts of the mesocarp. For *P. americana var. drymifolia*, the clean seeds were placed in 70% alcohol for 15 minutes and then planted in a laminar flow hood with superficial flaming. After that, part of the cotyledons was removed leaving a maximum of 1.5 cm^2 flanking the embryo. For *P. americana var. americana*, the rinsed seeds were immersed in a 10% sodium hypochlorite solution for 10 min, then placed in the 70% alcohol solution for later sowing in a laminar flow hood under the conditions just described.

2.3 Culture medium

MS culture medium was used [15], adding 30 g/L of sucrose (Bioxon: Cat. No. 217000), cytokinin (BAP) at 0.05 mg/L (Sigma: Cat. No. B3408), and 15 g/L of bacteriological agar (Bioxon: Cat. No. 215000). pH was adjusted to 5.7–5.8, and the medium was sterilized in an autoclave at 120°C at 20 kgf/cm^2 for 15 minutes.

2.4 *In vitro* germination conditions

Sowing was carried out in 105 × 55 mm glass flasks with 25 ml of the medium, in a laminar flow hood with 10 individuals of *P. americana var. drymifolia* and 10 of *P. americana var. americana*. Once sown, the seeds were maintained in a culture chamber at a temperature of 25 ± 1°C with a photoperiod of 16 light hours at an intensity of 80 E·m-2·s-1 against 8 hours of darkness.

Figure 1.
Characteristics of the avocado fruits in the mature consumption stage used in this work, Persea americana var. americana (A-C) and Persea americana var. drymifolia (D-F). The fruits were washed and separated into pericarp (A, D), mesocarp (B, E), and seeds (C, F). Scale line equivalent to 1 cm.

2.5 Determination of physio-morphological parameters

The seeds were evaluated at 0, 1, 3, 5, 10, 15, 30, and 60 days, using the first evaluations (0, 1, 3, 5, 10, and 15 days) to determine the days of rupture and emergence of the seedlings, while 30 and 60 days were selected to determine four key parameters of the two varieties: percentage of germination, contamination, oxidation, and death or necrosis. Growth parameters such as seedling size, root length, stem and root diameter, and dry weight.

2.6 Statistical analysis

All parametric tests were carried out using the Statistical Package for Social Sciences (IBM SPSS V.20). Mean differences were determined by a Tukey test with a significance value of $p < 0.05$ in 10 independent measurements.

3. Results

3.1 Characteristics of the fruits of *P. americana var. drymifolia* and *P. americana var. americana*

The fruits were weighed, measured, and separated into their tissues (pericarp, mesocarp, seeds) to determine proportions. The results for *P. americana var. drymifolia* were weights of 83.3–154.5 g with averages for the pericarp of 7.37 ± 1.43 g (7%), mesocarp of 73.09 ± 17.28 g (65%), and seeds of 31.32 ± 7.67 g (28%) (**Figure 2**). Seed size was 4.87 ± 0.49 cm. For *P. americana var. americana*, the fruits had weights of 332.4–481.5 g, with averages for the pericarp of 32.51 ± 6.16 g (8%), mesocarp of 316.46 ± 38.27 g (79%), and seeds of 50.06 ± 12.48 g (13%) (**Figure 2**). Seed size was 7.63 ± 0.15 cm.

3.2 *In vitro* establishment

The medium and conditions used to ensure asepsis of the seeds in this study allowed the germination of all the seeds sown. For *P. americana var. drymifolia*, the surface cleaning with 70% alcohol and superficial flaming were sufficient to lower the percentages of contamination, so that on the initial days of observation. and at the maximum time, homogeneity in seedling size and development of the leaves, stems,

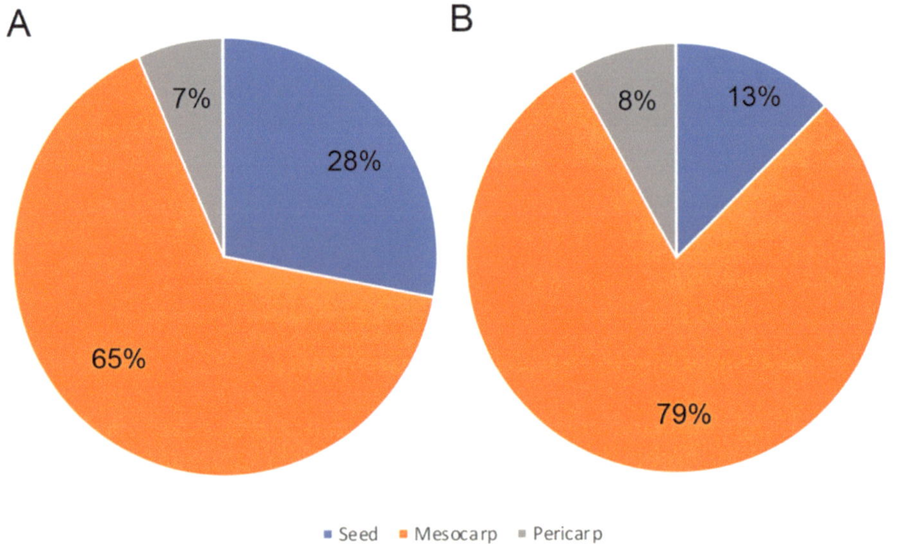

Figure 2.
Proportion of the avocado fruit tissues used in this study: A) Persea americana var. drymifolia; B) Persea americana var. americana.

Figure 3.
In vitro establishment of seeds of two avocado varieties, Persea americana var. drymifolia (A-E) and Persea americana var. americana (F-J). Embryos covered with 1.5 cm² of the cotyledon were planted in MS media (A, F). At 3 ± 1 days post-planting, the radicle emerged from the embryo (C, H). At 5 days, the emergence of the seedling is visible (B, G). At 10 and 30 days, development of the seedling and root is evident (D, E, I, J).

and roots was observed (**Figure 3**). For *P. americana var. americana*, high contamination percentages occurred due to the presence of endogenous bacteria, so a first cleaning step was carried out with 10% sodium hypochlorite to reduce contamination. Observations showed that homogeneity was achieved in the size and development of the seedlings (**Figure 3**).

3.3 Germination, oxidation, and contamination

The seeds of both varieties showed germination and seed breakage 3 ± 1 days after sowing. Determination of germination was made from the appearance of the radicle [16]. Emergence of the seedling was generally seen at five days, so the observations at 10, 15, and 30 days only served as controls for development and the possibility of loss due to contamination under the culture conditions used. The elimination of part of the cotyledon allowed an adequate exchange of water and nutrients in the germinative phase. The levels of this tissue allowed low oxidation and/or loss of the embryo that favored the development of the seedlings. The study obtained 100% germination and survival for both varieties, null oxidation levels, and only 5% contamination in the drymifolia variety (**Table 1**).

Avocado variety	GD	G	O	C	S
drymifolia	3 ± 1	100%	100%	5%	100%
americana	3 ± 1	100%	100%	0%	100%

GD.- Germination Day; G.- Germination; O.- Oxidation; C.- Contamination; S =.

Table 1.
Percentages of germination, oxidation, contamination, and seedling survival for seeds of two varieties of avocado, Persea americana var. drymifolia (Mexican native avocado) and Persea americana var. americana (west Indian avocado).

Avocado variety	SS (cm)	RL (cm)	SD (cm)	RD (cm)	DW (g)
drymifolia	56.94 ± 9.28	37.41 ± 8.9	0.29 ± 0.08*	0.23 ± 0.1*	0.52 ± 0.14
americana	54.28 ± 9.2	39.21 ± 7.67	0.31 ± 0.06*	0.25 ± 0.2*	0.54 ± 0.12

**: statistically-significant difference; SS.- Seedling size; RL.- Root length; SD.- Stem diameter; RD.- Root diameter and DW = Dry weight.*

Table 2.
Statistical significance in the comparison of the means of the growth variables for seedlings between the varieties Persea americana var. drymifolia (Mexican native avocado) and Persea americana var. americana (west Indian avocado) after 60 days.

Figure 4.
Development of avocado seedlings germinated from seed in an in vitro system using two varieties, Persea americana var. drymifolia (A), and Persea americana var. americana (B) 60 days after sowing. The seedlings show full development of their leaf, stem, and root tissues (B, D).

3.4 Survival and development of the avocado seedlings

The seedlings obtained from the *in vitro* germination process using the two varieties presented a homogeneity in size and development of their organs 60 days after sowing. Variation in the length of the aerial part and roots was not statistically significant between or within the varieties. Similarities in their dry weight were observed and quantified. However, statistically-significant differences were observed in the stem and root diameters, as these were slightly higher in the americana variety (**Table 2**). Regarding the proportion of the developed organs, observations showed that the root data presented the greatest inter- and intra-variety variation (**Figure 4**).

4. Discussion

Seeds transfer an organism's genetic information from one generation to another, thus allowing variability to develop in the species. Seeds in unfavorable conditions can enter dormancy, a survival mechanism in the presence of certain climatic conditions, such as very low temperatures, alternating dry and wet seasons, or desert climates [17]. On average, germination takes 20–75 days [18]. Elements such as the quality of light, or its absence, are other determining factors in the success of germination [19, 20]. The emergence of seedlings and their establishment are the key stages of development after germination and preconditions for successful establishment at the final site [21]. The controlled conditions of *in vitro* systems allow the germination and propagation of many plants in shorter periods [22]. In the case of avocado production systems, the commercial and nutritional importance [23] of the fruit and the need to obtain rootstocks suitable for the establishment of important cultivars –such as "Hass"– demand greater propagation of materials with desirable traits in a shorter time but with homogeneous development [24–26].

Various studies of avocado have sought to achieve the massive propagation of materials with outstanding characteristics, but numerous factors can affect success. Interaction with the environment, the type of plant material, and the developmental stage [27–39] are factors that can cause low spread rates. During traditional propagation, avocadoes may be exposed to attacks that can reduce germination, so a controlled germination system that optimizes plant development would seem to be ideal for producing and selecting plants of agronomic interest as future rootstocks. For avocado production, *in vitro* germination of the *P. americana* var. drymifolia and *P. americana* var. americana varieties are necessary for the medium- and long-term selection of plants with some degree of tolerance for biotic or abiotic stress.

The natural differences in the size and weight of avocado fruits have led producers to search for greater homogeneity in order to satisfy different market needs. The classification systems and taxonomy of the varieties of this fruit have multiple parameters used to identify the members of a specific botanical variety. For *P. americana* var. drymifolia, fruit weight averages 118–134 g in the ripe stage. Proportions of 82–94 g of the weight correspond to the pericarp-mesocarp, while 35–38 g correspond to the seed [40–42]. These data do not differ greatly from the figures obtained in this study, considering that the latter fall between the maximum and minimum values reported. For *P. americana* var. americana, fruit weights of 250–312 g have been reported, with pericarp-mesocarp proportions of 150–280 g and seed weight of 35-60 g [40].

Microbial contamination and darkening of tissues in *in vitro* systems, however, are two of the greatest challenges for avocado fruit [39]. Therefore, aseptic processes are a determining factor for the establishment and germination of seeds. Previous studies have reported protocols with up to 3% contamination in whole seeds and 5% oxidation in seeds from which part of the cotyledon was removed. As an oxidation inhibitor, researchers have added antioxidant compounds like thiosulfate and silver nitrate to the cultivation media [12]. The main causes of contamination in earlier studies involved bacteria contained in, or associated with, embryos. Under those conditions, low percentages of established embryos (up to just 60%) were achieved [33]. In the present study, in contrast, the percentages of success in the establishment of embryos and the elimination of contaminants (maximum 5% for *P. americana* var. drymifolia) are clearly superior to those previously reported (**Table 1**).

The percentages of *in vitro* germination in the avocado seedlings observed in this work (3 ± 1 day) also contrast with previous reports, which found variations in the composition and concentration of the components of the culture media, with germination times of 10–35 days [33, 40, 43], and 94–99% germination of the seeds, both whole and those that presented only partial sections of the cotyledon of up to 1.5 cm^2 covering the embryo [44] (**Table 1**).

Seedling emergence was observed with no differences days after sowing. Once again, this result contrasts with previous reports where emergence occurred after 15 days [12] The seedlings evaluated at 60 days of development showed a marked difference in size, as well, when compared to seedlings obtained from germinating embryonic axes [12]. Although there are significant differences in the stem and root diameters of the two avocado varieties, the production of cell mass is not reflected in the measurements of dry weight, suggesting that the increase in this diameter was due to the absorption of water from the culture media and/or to natural variations in the two varieties (**Table 2**).

The findings of this study show that the method applied contributed to obtaining avocado seedlings from seeds of the varieties analyzed that were free of pathogens and had clear homogeneity in their development. Therefore, this approach could be useful in avocado genetic improvement programs.

5. Conclusion

The germination of seeds of *P. americana* var. drymifolia and *P. americana* var. americana in an *in vitro* system made it possible to obtain homogeneous seedlings with complete tissue development that were free of pathogens. This system also improved germination time compared to *ex vitro* germination, so it could be useful in future research on avocado plants. The method may also prove useful for work on topics like genetic selection, environmental stress, and pathogen tolerance.

DOI: http://dx.doi.org/10.5772/intechopen.107005

Author details

Essoh Aimé Cesaire Elekou[1], Irene Perea-Arango[2], Luis María Suarez-Rodriguez[1] and Rodolfo López-Gómez[1*]

1 Instituto de Investigaciones Químico-Biológicas, Universidad Michoacana de San Nicolás de Hidalgo, Michoacán, Mexico

2 Centro de Investigación en Biotecnología (CEIB), Universidad Autónoma del Estado de Morelos, Mexico

*Address all correspondence to: rodolfo.lopez.gomez@umich.mx

References

[1] Suárez D, Melgarejo LM. Biología y germinación de semillas. In: Melgarejo LM, editor. Experimentos en Fisiología Vegetal. Colombia: Universidad Nacional de Colombia; 2010. pp. 13-24. Available from: https://www.uv.mx/personal/tcarmona/files/2019/02/Melgarejo-2010.pdf. ISBN: 978-958-719-668-9

[2] Williams LO. The botany of the avocado and its relatives. In: Sauls JW, Phillips RL, Jackson LK, editors. Proceedings of the International Tropical Fruit Short Course: The Avocado. Gainesville: Fruit Crops Dept., Florida Cooperative Extension Service. Institute of Food and Agricultural Sciences, University of Florida; 1976. pp. 9-15 http://www.avocadosource.com/Journals/ITFSC/PROC_1976_PG_9-15.pdf

[3] Raharjo SHT, Witjaksono NFN, Gómez-Lim MA, Padilla G, Litz RE. Recovery of avocado (*Persea americana* mill.) plants transformed with the antifungal plant defensin gene PDF1.2. *In Vitro* Cells Development Biology. 2008;**44**(4):254-262. DOI: 10.1007/s11627-008-9117-2

[4] Alcaraz AL. Biología reproductiva del Aguacate (Persea americana mill.). In: IMPLICACIONES Para LA OPTIMIZACIÓN del CUAJADO [Doctoral thesis. Estación Experimental La Mayora]. España: Universidad de Málaga; 2009 Digital file. Available from: https://docplayer.es/10953208-Tesis-doctoral-biologia-reproductiva-del-aguacate-persea-americana-mill-implicaciones-para-la-optimizacion-del-cuajado.html

[5] Jr KRJ. History, distribution and uses. In: Whiley AW, Schaffer B, Wolstenholme BN, editors. Avocado Botany Production and Uses. Wallingford Oxon: CAB International; 2002. pp. 1-14. DOI: 10.1079/9780851993577.0000

[6] Bergh B, Ellstrand N. Taxonomy of the avocado. California Avocado Society Year Book. 1986;**70**:135-146 http://www.avocadosource.com/cas_yearbooks/cas_70_1986/cas_1986_pg_135-145.pdf

[7] Montañez OI, Vargas SC, Cabezas GM, Cuervo AJ. Colonización micorricica en plantas de Aguacate (Persea americana L.). Revista U.D.C.A Actualidad & Divulgación Científica. 2010;**13**(2):51-60. DOI: 10.31910/rudca.v13.n2.2010.729

[8] Pliego-Alfaro F, Palomo-Ríos E, Mercado JA, Pliego C, Barceló-Muñoz A, López-Gómez R, et al. *Persea americana* avocado. In: Litz RE, Pliego-Alfaro F, Hormaza JI, editors. Biotechnology Fruit Nutritions Crops. 2nd Edition. Boston, USA: CABI; 2020:258-281. ISBN-13: 978 1 78064 829 3

[9] Ayala ST, Ledesma N. Avocado history, biodiversity and production. Sustainable Horticultural Systems: Issues, Technology and Innovation. Vol. 2. Springer; 2014. pp. 157- 205. DOI: 10.1007/978-3-319-06904-3_8

[10] Hiti-Bandaralage JCA, Hayward A, Mitter N. Micropropagation of avocado (Persea americana mill.). American Journal of Plant Sciences. 2017;**8**:2898-2921. DOI: 10.4236/ajps.2017.811197

[11] SINAREFI-SNICS-SAGARPA. Propagación de aguacate. 2012 https://www.gob.mx/cms/uploads/attachment/file/232194/Propagacion_de_aguacate.pdf

[12] Barrera-Guerra JL, Ramirez-Malagon R, Martinez-Jaime OA.

In vitro propagation of avocado (Persea drymifolia ness.). In: Proceedings of the Induced Mutations in Connection with Biotechnology for Crop Improvement of the FAO/IAEA. Peru, 5-9 October 1998: Division of Nuclear Techniques in Food and Agriculture; 1998. pp. 63-69 http://www-pub.iaea.org/MTCD/publications/PDF/te_1216_prn.pdf#page=69

[13] Ramírez-Guerrero LG, González-Rosas H, Calderón-Zavala G, Velázquez MJ, Cetina-Alcalá VM, Castillo-González AM, et al. Efecto de NaCl Y CaCl2 en el desarrollo de ejes embrionarios de Persea americana mill criollo y 'Hass' cultivados in vitro. Revista Chapingo, Serie Horticultura. 2010;**16**(3):161-167 https://www.redalyc.org/articulo.oa?id=60919865002

[14] López-Encina C, González-Padilla IM. A propósito de semillas. Enc. en la Biology. 1996;**33**:3 http://www.encuentros.uma.es/encuentros33 / semilla33.html

[15] Murashige T, Skoog F. A revised medium for rapid growth and bio assays with tobacco tissue cultures. Plant Physiology. 1962;**15**:473-497. DOI: 10.1111/j.1399-3054.1962.tb08052.x

[16] Finch-Savage WE, Leubner-Metzger G. Seed dormancy and the control of germination. New Phytologist. 2006;**171**(3):501-523. DOI: 10.1111/j.1469-8137.2006.01787.x

[17] Sanabria D, Silva R, Oliveros M and Manrique U. Germinación de semillas de las leguminosas arbustivas forrajeras Cratylia argentea y Cassia moschata sometidas a inmersión en ácido sulfúrico. Venezuela, Maturín: Instituto Nacional de Investigaciones Agrícolas; 2004. p. 7. https://www.redalyc.org/ articulo.oa?id=85716311

[18] Valio I. Inhibition of germination of coffee seeds (Coffea arabica L. cv. Mundo Novo) by the endocarp. Journal of Seed Technology 1980;**5**:32-39. https://www.jstor.org/stable/23432771

[19] Huxley P. Some factors which can regulate germination and influence viability of coffee seeds. Proceedings International Seed Testing Association. 1964;**29**:33-57

[20] Medina E. Introducción a la ecofisiología vegetal. Washington: Secretaría General de la Organización de los Estados Unidos Americanos; 1977. https://www.uv.mx/personal/tcarmona/files/2019/02/Medina-1977.pdf

[21] James JJ, Svejcar TJ, Rinella MJ. Demographic processes limiting seedling recruitment in arid grassland restoration. Journal of Applied Ecology. 2011;**48**:961-969. DOI: 10.1111/j.1365-2664.2011.02009.x

[22] Kumar PP, Loh CS. Plant tissue culture for biotechnology. In: Altman A, Hasegawa PM, editors. Plant Biotechnology and Agriculture. Prospects for the 21st Century. London, UK: Academic Press, Elsevier; 2012. pp. 131-138. DOI: 10.1016/B978-0-12-381466-1.00009-2

[23] Ben-Ya ́acov, A. and Michelson, E. Avocado Rootstocks. Horticultural Reviews. Vol. 17. New York: John Wiley & Sons, Inc.; 1995. pp. 381-429. DOI: 10.1002/9780470650585.ch11

[24] Ben-Ya ́acov, A. Avocado rootstock-scion relationships. South Afric. Avocado Growers Association Yearbook. 1987;**10**:30-32 http://www.avocadosource.com/wac1/wac1-p030.pdf

[25] Pliego-Alfaro F, Bergh BO. Avocado. In: Hammerschlag FA, Litz RE, editors.

Biotechnology of Perennial Fruit Crops. Wallingford: CABI International; 1992. pp. 323-333

[26] Litz RE, SHT R, Gómez-Lim MA. Avocado. In: Pua EC, Davey MR, editors. Transgenic Crops V. Biotechnology in Agriculture and Forestry. Berlin: Springer; 2007. pp. 167-187

[27] Cooper PA. Advantages in the micropropagation of avocado (Persea americana mill). Acta Horticulturae. 1987;**212**(2):571-575. DOI: 10.17660/ActaHortic.1987.212.92

[28] Dalsaso L, Guevara E. Multiplicación clonal in vitro del aguacate (Persea americana) cv. "Fuerte". Agronomía Costarricense. 1988;**13**(1):61-71. http://www.mag.go.cr/rev_agr/v13n01_061.pdf

[29] Zirari A, Lionakis SM. Effect of cultivar, explant type, etiolation pretreatment and the age of plant material on the in vitro regeneration ability of avocado (Persea americana). Acta Horticulturae. 1994;**365**:69-76. DOI: 10.17660/ActaHortic.1994.365.6

[30] Castro M, Oyanedel E and Cautin R. In vitro shoot proliferation in avocado (Persea americana mill) induced by CPPU. In: Proceedings of the World Avocado Congress III. 1995. pp. 223-226. http://209.143.153.251/WAC3 / wac3_p223.pdf

[31] Barringer SA, Yasseen YM, Splittstoesser EW. In vitro multiplication and plantlet establishment of avocado. In vitro Cellular Development Biology Plant. 1996;**32**(2):119-121. DOI: 10.1007/BF02823142

[32] Barceló-Muñoz A, López-Encina C, Simón-Pérez E, Pliego-Alfaro F. Micropropagation of adult avocado. Plant Cell, Tissue and Organ Culture. 1999;**58**:11-17. DOI: 10.1023/A:1006305716426

[33] Rodríguez NN, Capote M, Zamora V. Cultivo in vitro del aguacatero (Persea americana Mill). Revista Chapingo Serie Horticultura. 1999;**5**:231-237 http://www.avocadosource.com/wac4/wac4_p231.pdf

[34] De la Viña G, Barceló-Muñoz A, Pliego-Alfaro F. Effect of culture media and irradiance level on growth and morphology of Persea americana mill microcuttings. Plant Cell, Tissue and Organ Culture. 2001;**65**:229-237. DOI: 10.1023/A:1010675326271

[35] Vidales-Fernández I. Efecto de los reguladores de crecimiento en los procesos de organogénesis y embriogénesis somática de aguacate (Persea americana Mill.). Thesis. Tecomán, Colima, México: Universidad de Colima; 2002. pp. 62-119

[36] Barceló-Muñoz A, Pliego-Alfaro F. Micropropagation of avocado (*Persea americana* mill). In: Jain SM, Ishii K, editors. Micropropagation of Woody Trees and Fruits. Forestry Sciences. Vol. 75. Dordrecht: Springer; 2003. pp. 519-542. DOI: 10.1007/978-94-010-0125-0_17

[37] Nhut DT, Thi NN, Khiet BLT, Luan VQ. Peptone stimulates in vitro shoot and root regeneration of avocado (Persea americana mill). Scientia Horticulturae. 2008;**115**:124-128. DOI: 10.1016/j.scienta.2007.08.011

[38] Zulfiqar B, Abbasi NA, Ahmad T, Hafiz IA. Effect of explant sources and different concentrations of plant growth regulators on in vitro shoot proliferation and rooting of avocado (Persea americana mill.) cv. "Fuerte". Pakistan Journal of Botany. 2009;**41**(5):2333-2346. http://www.pakbs.org/pjbot/PDFs/41(5)/PJB41(5)2333.pdf

[39] Cortés-Rodríguez MA, López-Gómez R, Martínez-Pacheco MM,

Suárez-Rodríguez LM, Hernández-García A, Salgado-Garciglia R. In vitro propagation of Mexican race avocado (Persea americana Mill var. drymifolia). Acta Horticulture. 2011;**923**:47-52. DOI: 10.17660/ActaHortic.2011.923.5

[40] Sánchez-Pérez JDLL. Recursos genéticos de Aguacate (*Persea americana* mill.) Y especies afines en México. Revista Chapingo, Serie Horticultura. 1999;**5**:7-18. http://www.avocadosource.com/wac4/wac4_p007.pdf

[41] Acosta DE, Almeyda LIH, Hernández TI. Evaluación de aguacates criollos en Nuevo León, México: región norte. Revista Mexicana de Ciencias Agrícolas. 2013;**4**(4):531-542 http://www.scielo.org.mx/scielo.php?script=sci_arttext&pid=S2007-09342013000400004

[42] Montes-Hernández S, De la Torre-Vizcaino JD, Heredia-García E, Hernández-Martínez M, Camarena-Hernández MG. Caracterización morfológica de germoplasma de aguacate Mexicano (Persea americana var. drymifolia, Lauraceae). Interciencia. 2017;**42**(3):175-180 https://www.redalyc.org/journal/339/33950011006/html/

[43] Sánchez-Romero C, Perán-Quesada R, Márquez-Martín B, Barceló-Muñoz A, Pliego-Alfaro F. In vitro rescue of immature avocado (*Persea americana* Mill.) embryos. Scientia Horticulturae. 2007;**111**:365-370. DOI: 10.1016/j.scienta.2006.11.009

[44] Capera E, Godoy JV, Amórtegui I. El cultivo de aguacate: Módulo educativo para el desarrollo tecnológico de la comunidad rural. Repositorio Institucional Agrosavia. Corporación Colombiana de Investigación Agropecuaria. 2001. p. 49. Web. Available from: http://bibliotecadigital.agronet.gov.co/bitstream/11348/4911/1/El%20cultivo%20del%20aguacate.pdf

Chapter 6

Influence of Abiotic Stresses on Seed Production and Quality

Rupa Das and Saikat Biswas

Abstract

Climate change is exerting detrimental impacts on agriculture through various biotic and abiotic stresses. Abiotic stresses such as drought, flood, temperature extremes, salinity, chemicals, heavy metals, nutrient scarcity/toxicity, wind and light in combination more adversely affect the seed production and quality by hampering plant's morphological, physiological, cellular, biochemical and molecular activities than alone, resulting in poor production of high-quality seeds. Deterioration of yield and quality arises also under abiotic stresses. Under abiotic stresses, plant activates its own defensive mechanisms by escaping, avoiding and tolerating stresses. Some of the plant's defensive mechanisms include plant's morphological, cellular, physiological, biochemical and molecular changes to adapt the stresses, synthesis of compounds such as ABA, proline, polyamines increasing the activities of ROS quenchers, expression of stress-resisting genes and activation of enzymes. Further, exogenous application of phytohormones, stress-alleviating compounds, modification of agronomic management, modern breeding strategies such as development of resistant varieties can also help to cope up with stresses and produce quality seeds. Financial and policy support of government or NGOs regarding development of infrastructure, research technologies and thereby, multi-locational trials as well as technology transfusion through extension activities are needed to curtail down the devastating impact of abiotic stresses on quality seed production.

Keywords: abiotic stresses, agronomic management, climate change, defensive mechanisms, seed production, quality

1. Introduction

Improving agricultural productivity is the prime goal of farmers around the globe to meet the food demand of the entire population. Over the years, the global population is increasing at alarming rate and likely to increase even further in future. Food scarcity is therefore a major issue which many countries of the world are experiencing now. Rapid industrialization/urbanization, migration of rural people to urban areas for alternative job opportunities, literacy, low agricultural production, lack of policies etc. are key reasons for which agriculture sector is suffering currently. In order to achieve the food security, agricultural production needs to be lifted at this hour. However, shrinkage of agricultural land in resonance with population growth forces the farmers to go for intensive farming approach. In intensive farming, excessive uses of chemicals and unscientific management practices not only degrade the soil, water and other precious natural resources of the crop production, but also create climate change issue which possess a major threat to sustainable agricultural production. Further, industrialization/urbanization, deforestation or any other anti-nature practices aggravate this climate change situation and thereby, affect crop production.

Performance of a crop relies on the interaction of its genetic traits with the growing environment. Crop plants suffer from climate change scenario due to their sessile growth habit. Climate change exerts various stresses on crop plants and thereby, affects the crop negatively in most cases. Such stresses can be exhibited from the living (biotic) or non-living (abiotic) things of the environment. In nature, various stresses usually put combined impact on the crop plants. Abiotic stresses such as drought, flood, salinity, temperature extremes (hot/cold), heavy metals, light, wind, nutrients/chemicals etc. on plants are either physical or chemical in nature [1]. These abiotic stresses resulted from climate change determine the distribution of plants in different environmental conditions [2] and thereby, influence the crop productivity throughout the globe [3]. These stresses during any crop growth stages specially, reproductive stage markedly reduce yield of the crop by affecting reproductive organs [4]. They are also known to influence biotic stresses positively, which in turn, affect agricultural production through disruption of seed germination, vegetative growth, dry matter production and its partitioning to reproductive parts [5]. Around 70% of yield reduction in the world is due to abiotic stress [6]. Severe stresses not only affect the yield but also seriously hamper the quality of the produce through creating imbalance between demand and supply of nutrients, inactivation of enzymatic activities, suppression of various genes responsible for the quality expression etc. [7]. As these abiotic stresses put barriers on global food security, it is needed to develop suitable strategies to achieve considerable amount of yield with maintained quality under climate change situation.

In this context, seed is a key driver of agriculture. Quality seed production and its utilization can play major roles in achieving high yield under climate change scenario. However, quality seed production under various abiotic stresses needs special emphasis as these stresses exert detrimental effects on plant. Abiotic stresses limit the production and quality by affecting the crop plants from morphological to molecular levels [8]. However, in response to the stresses, plant shows few mechanisms like escaping stress, stress avoidance and stress tolerance. Plant undergoes certain molecular, cellular and physiological changes to cope up with stresses and acclimatizes under stress situation. Insight view to such changes is one of the fundamental components of plant studies to ensure food security. Improved understanding of plant response to stresses can help to develop various strategies to alleviate the stresses. Such strategies include various traditional and modern breeding, agronomic management practices along with exogenous applications of stress tolerating compounds to cope up with abiotic stresses. This chapter highlights the impacts of abiotic stresses on crop plants specially seed production and quality as well as points out various agronomic management strategies to cope up with the abiotic stresses by providing the understanding of plant's defense against such stresses.

2. Stress

Stress, in general, refers to any deviation from the normal condition of a character or phenomena. In biology, stress mostly exerts adverse impact on individual or group or mass. In agriculture, negative impact of stress is seen on crop growth, development, productivity and quality. Stress induces various responses inside the plant such as change in gene expression, change in cell metabolic activities, and changes in plant's physiological and biochemical activities and thus, affects its growth rate, productivity and quality. In environment, stress is of two types: biotic stress and abiotic stress (**Figure 1**). Biotic stress arises due to interaction between living organisms, while abiotic stress is a result of interaction between organisms with non-living environmental properties. In reality, both these two stresses are linked to each other.

2.1 Biotic stress

Biotic stress arises when living organisms such as weeds, insects, diseases (caused from bacteria, fungi, virus etc.) exert stresses on plant and thereby, hampers its growth and development. The kinds of biotic stress on plant depend on edaphic and climatic conditions. However, yield loss is a common occurrence due to this stress.

2.2 Abiotic stress

Physical or chemical environment puts adverse effect on the living organisms and thus, creates abiotic stress. In agriculture, abiotic stress results in impairment of normal growth and development of plants and thereby, crop yield, quality and farmers' income get hampered. It is further noted that in environment, several stresses together put combined effect on plant and thereby, deteriorate crop productivity and quality in greater extent as compared to their individual effect. It adversely affects the crop from morphological to physiological, biochemical or molecular levels. Abiotic stress includes water stress (drought and flood), salinity stress, temperature stress

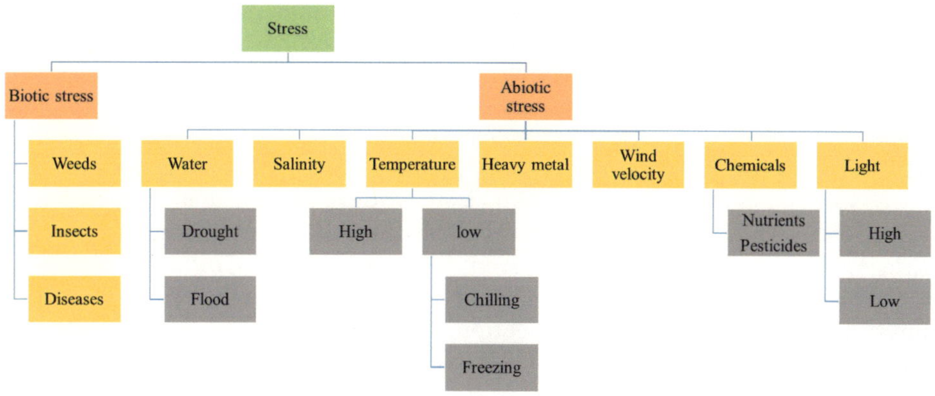

Figure 1.
Various types of stress.

(heat and cold), heavy metal toxicity, nutrients and pesticides toxicities, light stress (high and low). Some other types of abiotic stresses are shade, UV exposure, photo-inhibition, air pollution, wind velocity etc. In following section, descriptions of these stresses are briefly mentioned.

2.2.1 Water stress

Water is the precious input for any life and obviously, for the crop. Plant's photosynthesis, transpiration, nutrient uptakes, translocation of assimilates etc. depend on water and thereby, its imbalance in plant can cause serious damage to plant. The uneven distribution of water is resulted from climate change scenario which the globe is experiencing now. Generally, water stress occurs in the form of drought and flood.

Drought: In meteorology, drought is defined as the period of deficiency of rainfall from the normal in an area. In hydrology, drought arises from low river and stream flows and reduced ground water table. In agriculture, drought indicates the dry period during the crop growth period which critically hampers the crop growth and yield. Sometimes, plant is unable to uptake soil moisture even after its availability due to high salt concentrations in soil solution. This phenomenon is known as apparent drought. Drought can also be resulted from high and low temperature. High temperature induces evapotranspiration loss of water and thus, creates drought like situation. In case of cold condition like at freezing temperature, ice crystals are formed in the extracellular spaces of plant, resulting in loss of the water potential which finally causes intracellular water efflux. Drought stress can lead to loss of plant vitality and alter in plant's normal functioning.

Flood: Flood is the opposite situation of drought. It is the situation of excess water in an area for a period of time. Flood when occurs from the sudden outburst of cloud and results in excessive rainfall in a short spell of time is known as flash flood. Flash flood lasts for a short period of time generally up to few weeks. Another type of flood is deep water flood which last for a longer period of time.

2.2.2 Salinity stress

Salinity is a major problem in any part of the world especially in arid and semi-arid areas where potential evapotranspiration is higher than the rainfall and poor leaching of salts beyond root zone is seen due to insufficient rainfall. Salinity can be defined as presence of excess quantity of salts in the soil, which negatively impacts on the crop [9]. Salinity is measured using electrical conductivity (EC). In general, soil having EC >4 dS/m, exchangeable sodium percentage (ESP) < 15.0 and pH. <8.5 is considered as saline soil [10]. Saline soil contains chloride, sulfate salts of sodium, magnesium and calcium ions. Salinity in soil can be developed by both natural and anthropogenic means. Weathering of rocks, flooding and intrusion of sea water to agricultural land, seepage of saline water, wind blow etc. are the natural cause of build-up of salts in the soil. Human induced causes of soil salinity are poor water quality of irrigation, deforestation, overgrazing, intensive cropping etc. Excessive soil salinity leads to deterioration of soil health through changing cation exchange capacity, damaging soil physical structure through deflocculation and reduction of hydraulic conductivity and hampering soil microbial activity.

2.2.3 Temperature stress

Abrupt change in temperature in response to global climate change is a great concern. Increase of temperature due to global warming and anthropogenic causes can change biodiversity, crop ecosystem and thus, limit the crop production to a high extent. High temperature or heat stress induces higher respiration over photosynthesis and thereby, leads to starvation injury through loss of food reserves in plants. Plants are differentiated based on their degree of high temperature tolerance: psychrophiles (up to 15–20°C), mesophiles (up to 35–45°C) and thermophiles (up to 45–100°C) [11]. Low temperature or cold stress is the opposite phenomena of heat stress and is mostly experienced in temperate areas. It is of two types: chilling stress and freezing stress. Chilling and freezing stresses affect the crop's physiological, biochemical functions and thereby, hamper its growth and productivity. According to the tolerance to cold temperature, plants are classified into 3 categories: Chilling sensitive (Plants are extremely sensitive above 0°C and below 15°C), chilling resistant (plants tolerate low temperature but adversely affected under formation of ice crystals in extracellular spaces) and frost resistant (plants tolerate extremely low temperature).

2.2.4 Heavy metal toxicity

Heavy metals are inorganic, non-biodegradable compounds having atomic mass and density > 20 and > 5 g/cm^3, respectively. They cause mutagenic effects on plants through contaminating irrigation, food chain and environment [12]. Application of excessive quantity of water from contaminated source and fertilizers/pesticides results in accumulation of heavy metals in soil and from soil, plant absorbs them. In general, heavy metals (Ag, Cr, Cd, As, Sb, Pb, Se, and Hg) are non-essential and at higher concentrations can affect plant's normal functioning. Besides, some essential elements such as Zn, Cu, Ni, Fe, Co etc. at higher concentrations pose heavy metal toxicity in plant.

2.2.5 Chemical toxicity

Consistent reliance of agriculture on inorganic chemical fertilizers and pesticides (herbicides, insecticides, fungicides etc.) from green revolution is increasing day by day. Besides, rapid industrialization and use of sewage water without treatment possess threats to the crop through the adverse effects of the chemical toxicity on the soil- plant system.

2.2.6 Nutrient toxicity/deficiency

Excessive application of fertilizers can lead to toxicity of nutrients in the soil. On the other hand, scarce application of nutrients leads to deficiency of nutrients.

2.2.7 Wind velocity

High wind velocity over the cropped area can put stress on the growth and development of the crop. Wind velocity is resulted from the movement of wind from one direction to another at certain speed.

2.2.8 Light stress

Light is essential input for plant growth. In fact, crop production is synonymous with harvesting solar energy. When light becomes excessive or insufficient, it causes detrimental effect on the plant. This phenomenon is known as light stress.

3. Effect of abiotic stresses on plants

Abiotic stresses alone or in combination influence negatively on plants (**Figure 2**). It has been observed in various regions and crops that there is generation of H_2O_2 under abiotic stress. Closure of stomata partially or completely under abiotic stress such as drought stress reduces photosynthesis by restricting entry of CO_2 and hampering electron flow through electron transport chain. Under abiotic stress, further, hydroxyl radicals, superoxide radicals and singlet oxygen are formed. These altogether adversely affect lipid (lipid peroxidation), protein (oxidation), nucleic acids and enzyme activity leading to cell death. Some adverse effects of different abiotic stresses on plant are briefly discussed below.

3.1 Water stress

Drought: Under scarcity of water, the seed germination and early seedling establishment get hampered due to depletion of seed reserves as well as mechanical obstruction made by the hard soil, which are followed by restricted vegetative growth.

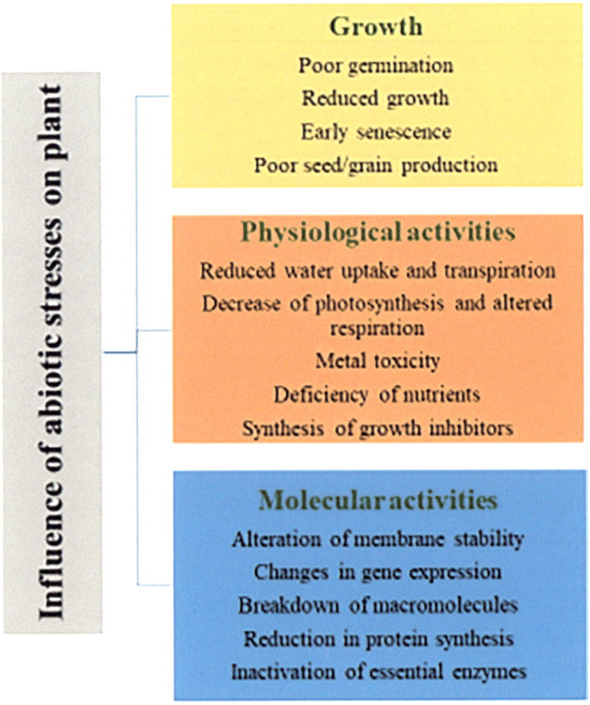

Figure 2.
Influence of abiotic stresses on plant.

Some other major impact of drought includes reductions of leaf area, chlorophyll a, b, carotenoids, stomatal conductance, rubisco activity leading to poor photosynthesis and translocation of photo-assimilates from source to sink. It impairs photosystem (PS) I and II as well as reduces starch biosynthesis through hampering ribulose phosphatase activity. Consequently, dry matter production and development of reproductive structures are negatively impacted. Under drought stress, concentrations of cell solutes increase and pose toxic effects on plants. This stress further increases the intra- and inter-competitions for water among crop plants and between crop and weeds. It increases sucrose viscosity leading to restricted movement of sap inside plant. Under drought stress availability of nutrients are variably altered. For instance, availability of nitrogen increases, while phosphorus availability decrease and no distinct effect is seen on potassium under drought stress in root vicinity, resulting in variability in uptake by the plant and consequently, metabolism of nutrients in cell is affected [13]. Assimilation of ammonia to organic form is restricted as activities of nitrate reductase, glutamine synthetase etc. are reduced. C_4 plants are more sensitive to drought stress than C_3 plants due to closure of stomata which ultimately reduces photosynthesis [14]. Drought restricts mineral uptakes and nitrogen fixation ability in various leguminous crops.

Flood: Flood generally creates deficiency of oxygen/hypoxia as anaerobic situation is formed through either water logging or submergence. Flood reduces the movement of oxygen and other gases in root zone of plant and ethylene diffusion from plant, which causes chlorosis of plant leaves. Negative impact on plants due to flood includes wilting of shoot, loss of chlorophyll pigments, decay and death of leaves/aerial parts, abscission, epinasty, lenticel formation, build-up of toxins under hypoxic environment, less root respiration, root proliferation and other physiological disorders.

3.2 Salinity stress

Salinity stress possess two major primary effects on plants: osmotic stress and ion toxicity. As already mentioned earlier, this stress can induce the drought stress by restricting the uptake of water by plants from the soil through exerting higher osmotic pressure to root cell (osmotic pressure of soil solution> osmotic pressure inside plant cell sap). Even if water uptake occurs, water also carries lots of salts (Na^+, Cl^- etc.) inside the plant along with it and these excess salt ions can pose injury at the cellular level of the plant by hampering some essential enzyme activities. Production of reactive oxygen species (ROS) owing from oxidative stress results in detrimental effect on protein, nucleic acid and certain enzymes of plant [15]. Under excess salts, plant can show burnt like appearance in leaves and experiences deficiency of some essential elements such as calcium (Ca^{2+}), potassium (K^+), magnesium (Mg^{2+}), nitrate (NO_3^-) etc. and abundance of Na^+ near root zone. Calcium (Ca^{2+}), potassium (K^+), magnesium (Mg^{2+}) and nitrate (NO_3^-) play major role in photosynthesis and less uptake due to salinity leads to reduction of photosynthesis and translocation of assimilates from source to sink. It causes poor leaf expansion, stunted growth and less dry weight of plant, sterility of florets and loss of pollen viability in plant. Under excessive salt situation, there is an increase in epidermal thickness, mesophyll thickness, palisade cell length and diameter and spongy cell diameter and reduction of intercellular space in leaves. Salinity stress specifically, increases Na^+/K^+ ratio of the cell. Imbalance supply of potassium reduces cell turgidity, enzyme activity and membrane potential of plant. It further induces some secondary effects on plants such as reductions of cell expansion as well as division, membrane stability and cytosolic metabolism as dominance of Na^+ in cell creates restriction of some essential enzyme activities. Under salinity stress, closure of stomata partially or completely reduces transpiration and cell division. Consequently, there is an increased rate of reduction of plant growth, high degree of defoliation as well as senescence and plant's death [16].

3.3 Temperature stress

High temperature/heat stress: Heat stress often increases the evapotranspiration loss of water and thereby, creates drought like situation. This situation is very critical when soil temperature increases along with drought stress. Elevation in temperature causes denaturation of protein, inactivation of essential enzyme activities (*viz*., sucrose phosphate synthase, adenosine diphosphate-glucose pyro-phosphorylase, invertase etc.) reductions of starch and sucrose synthesis [17], generation of ROS and death of the cell [18]. Further, light dependent reactions in stroma of thylakoid and carbon metabolism are hampered under elevated temperature. Impairment of electron supply to PS II under heat stress leads to restriction of the activity of PS II. Seed germination and seedling stand establishment are greatly affected by this stress. At high temperature, respiration exceeds photosynthesis and high loss of carbon through photo-respiration is seen (respiration rate increases two times with each 10°C increase of tissue temperature). Root number, length and biomass are greatly affected under high temperature rise, resulting in decreased uptakes of nutrients and water from soil. Further, drying or scorching of aerial part of the plants, reduction of plant growth and development through depleting chlorophyll as well as hampering photosynthetic activity and cell division are witnessed under heat stress. It is noted that sudden increase in temperature is comparatively more harmful as compared to gradual temperature rise as it disrupts biochemical, physiological and molecular activities of the plant. C_3 plants suffer comparatively more under heat stress over C_4 plants due to variations in energy supply and carbon metabolisms [19].

Low temperature/cold stress: Cold stress causes chilling (dysfunctioning of physiological properties) and freezing (dehydrating the cell) injuries to the plant. Low temperature stress hampers membrane integrity resulting in electrolyte leakage/ plasmolysis, alteration of cell metabolic activities and reduction of protoplasmic streaming [20]. It alters nucleic acid and protein synthesis and down-regulates PS II [21]. Wilting, bleaching through pigment photo-oxidation, leaf necrosis, browning, cell death etc. are the common examples of impact of cold stress on plant.

3.4 Heavy metal toxicity

Heavy metals adversely affect plant's physiological, morphological, biochemical and molecular activities. After root uptake, heavy metals *viz*., Pb, Cu, Hg, etc. translocate inside the plant through xylem in response of transpiration pool and affect nutrient distribution, photosynthesis, enzyme activities, Cu/ Zn-SOD, ethylene receptors etc. and consequently, decrease molecular oxygen and synthesize reactive oxygen species (ROS) (oxidative stress) [22]. Formation of ROS damages the cell further.

3.5 Other abiotic stresses

Excessive and pointless application of chemicals can lead to pollution of the environment and damages ecosystem health and diversity. Further, it induces climate change which in turn, affects the crop growth, yield and quality. Use of excessive nutrients from inorganic sources also play similar impact on environment. These chemicals, fertilizers persist in nature and create toxic effects on crop plant. For instance, reliance on chemical based farming activities can degrade the soil health and reduces the soil productivity. Ground water, surface water contamination, air pollution etc. are now common phenomenon which aggravate climate change scenario. Plant previously suitable for a region is now therefore facing trouble under such changing climate. It induces various other abiotic stresses and biotic stresses to negatively impact on crop's germination, photosynthesis, translocation of assimilates etc. and thereby, hampers crop yield. On the other hand, scarce nutrients in the soil can lead to nutrient deficiency symptoms on plant through less uptake and thereby, affects plant growth, development, yield and quality as these nutrients are essential for the plant's nourishment.

High wind velocity leads to lodging situation in tall growing plant and thereby, affects their growth. It further creates soil erosion and thus impact negatively by washing away the nutrients essential for crop growth. Further, under high wind velocity, evapotranspiration loss of water is increased which creates scarcity of water for the plant growth.

Light plays an important role in photosynthesis. Exposure to less light can lead to less photosynthesis. On the other hand, excessive light can damage photosynthetic apparatus (photoinactivation and photodamage). Low light or shading affects the plant growth and development. High light intensity can lead to heat or drought stress like situation and thereby, affects the plant growth. UV ray damages DNA and causes leaf bleaching, oxidative stress by formation of ROS. Under high light, plant experiences breakdown of D1 protein of PS II and reduction of PS I polypeptides such as PsaA, PsaB and PsaC [23].

4. Influence of abiotic stresses on seed production and quality

Abiotic stresses exert negative influence on production of seed primarily by hampering germination as well as photosynthetic efficiency and translocation of assimilates from source (leaf) to sink (seed). Further, flowering, fertilization and seed filling phases are greatly affected under abiotic stresses. Reduced activity of acid invertase restricts reproductive parts of plant through phloem unloading. Reduced endosperm size, abortion of embryo, less seed filling, unfilled seed etc. drastically reduce the seed production and quality. Deterioration of yield and quality arises also from deficiency of supply of essential inputs to the plant under abiotic stress. The influence on seed formation and quality is resulted from complex relationship between individual stresses. For instances, changes in gene expression, oxidative damage, alteration of ion distribution and cell homeostasis lead to loss of production and quality of crop under abiotic stresses. Reduced productions of various crops under abiotic stresses have been listed in **Table 1**. The specific influence of different abiotic stresses on seed production and quality are briefly discussed here under.

4.1 Water stress

Drought is the principal limiting factor of growth, yield and quality of cereal crops throughout the world. Drought (water deficit) affects the plant at all the stages by altering its physiological, biochemical, molecular properties. Degree of yield loss depends

Crop	Abiotic stress	References	Crop	Abiotic stress	References
Lentil	Drought stress	[24]	Soybean	Cold stress	[25]
Chick pea		[26]	Rice		[27]
Soybean		[28]	Pea		[25]
Common bean		[29]	Chick pea		[25]
Mung bean		[30]	Broad bean		[25]
Faba bean		[31]	Soybean	Salinity stress	[32]
Barley		[7]	Chick pea		[32]
Wheat		[33]	Lentil		[32]
Cotton		[34]	Mung bean		[35]
Maize		[36]	Faba bean		[37]
Spotted bean		[24]	Wheat		[38]
Black gram		[39]	Groundnut, Chickpea, Green gram, Soybean, Pigeon pea	Cd stress	[32]
Cow pea		[40]	Grass pea, Chick pea	Pb stress	[32]
Pigeon pea		[31]	Chick pea, Green gram	Cr stress	[32]
Lupin		[31]	Pea, Lentil, Soybean, Black gram	Hg stress	[32]
Lentil	Heat stress	[41]	Pea, Chick pea, Cowpea, Green gram	Cu stress	[32]
Wheat		[42–45]	Chick pea, Cowpea, Pigeon pea	Ni stress	[32]
Rice		[46]	Cowpea, Chick pea	Zn stress	[32]
Ground nut		[41]	Pea, Chick pea	As stress	[32]
Chick pea		[41]			
Pea		[47]			
Pigeon pea		[41]			
Cow pea		[41]			
Soybean		[41]			
Mung bean		[47]			
Common bean		[25]			
Broad bean		[47]			
Lupin		[47]			

Table 1.
Major crops under harmful impact of abiotic stresses on production.

on occurrence and severity of drought. However, drought stress at flowering as well as post flowering grain filling period is extremely detrimental which reduces the seed yield most. Drought reduces production of flowers, pollen viability, fertilization and seed filling, resulting in loss of production qualitatively and quantitatively. Shortening of pollination and seed filling periods and early maturity lead to less or immature production of seeds. Abortion of reproductive parts of plant is a common phenomenon under drought stress at reproductive stage. Further, under drought stress, increment of toxic ion concentrations in cell, loss of cell turgidity through disruption of water streaming etc. reduce leaf growth leading to poor photosynthesis as well as translocation of dry matter from source to seed (sink). In legume crops, drought stress reduces nodulation, nitrogen fixation and thereby, hampers seed production. Few research works [7] revealed that mild drought stress during seed filling stage leads to increase in protein content of plant. For instance, [48] reported loss of gluten, glutenin, gliadin etc. but increase of protein in wheat under drought stress. However, severe water deficit leads to reduction in amino acid pool and its incorporation to protein as well as less accumulation of N, P, Fe, and Zn, which in turn decreases the protein content of the seed. It has been also observed from many research work that protein content increases after withdrawal of drought in seed filling stage [49]. Oil, oleic acid, glucose, sucrose, fructose etc. decrease with increase of drought during seed filling stage. PUFA contents of sunflower [50] and groundnut [49] are found to decrease under drought stress. β-glucan content of seed also decreases under drought stress. Increased sucrose content and less formation of starch in potato are observed under drought stress [51]. Reduction of starch synthetase activity under drought stress leads to amylose content of wheat [52]. Further, drought stress reduces uptake of nutrients and thereby, lowers the nutritional quality of seed. Increase of electrical conductivity, poor germination and vigor can be seen from the seeds produced under drought stress.

Excessive water use can lead to deficiency of oxygen and accumulation of toxic chemicals inside the plants and thus, hampers number of spike/pod per plant, number of seeds per spike or pod, seed weight leading to reduction of seed production. Further, restriction in uptake of nutrients specially, N under high water content due to leaching and/or dentrification losses can result in quality deterioration of the produce, specially, protein.

4.2 Salinity stress

Salinity is especially harmful at seedling, flowering and reproductive stages. It plays detrimental role in inhibiting seed germination or causing seedling mortality and less plant population which ultimately affects the seed yield. Further, production of seeds are greatly affected under salinity stress due to reductions in photosynthesis, transpiration, stomatal conductance and metabolic activities owing from less chlorophyll, carotenoid, relative water content (RWC) and increased ion toxicity. Nodulation and nitrogen fixation in leguminous crops are also affected by excess salt content. Poor pollen viability and restricted translocation of assimilates to seed lead to low seed production. Restricted supply of nutrient and water to the plant from soil results in detrimental influence on seed production and quality of crops. Oil, protein and starch contents decrease due to disturbances in nitrate uptake and nitrogen metabolism under high salinity. PUFA content of sunflower decreases under drought stress [53].

4.3 Temperature stress

Temperature stress (high/low) greatly affects the seed production and quality of the produce. Loss of yield is resulted from poor seed germination and plant stand establishment. Further, high temperature during flowering and grain filling period can lead to pollen desiccation or sterility, fertilization failure, dysfunctioning of tapetal cells, loss of cell turgor, less leaf area (and increased senescence rate), reductions of chlorophyll, CO_2 assimilation or photosynthesis and increment of photorespiration [54]. Reduction of net photosynthesis due to poor rubisco activity and decrease in translocation of assimilates hamper the production of seed. It has been also noticed that heat stress reduces seed filling duration resulting in loss of production as well as its quality [55]. Increment of soil temperature hampers root growth and increase respiration rate, which further restrict uptake of nutrients and water from soil resulting in not only loss of yield but also changes in quality of the produce. Heat stress, specifically, restricts starch and protein synthesis, which are major determinants of seed quality. For instances, during seed filling period, increase of temperature leads to imbalance supply of nitrogen in pulses [56]. Starch and protein contents also decline under heat stress in wheat and maize [57, 58]. However, it has been also noticed that heat stress has positively affected on seed protein content. For instances, elevated soil temperature (around 2.5°C) has significantly improved protein content (aspartate, glycine, alanine, arginine, valine and tryptophan) of barley [59]. Lin et al. [60] observed that increased temperature during early seed filling period improved protein content in rice but reduced prolamins. In leguminous crops, heat stress mostly reduces protein contents and increases oil contents due to antagonistic effect between oil and protein [61]. Heat stress during flowering and seed filling stages reduces the starch synthesis period resulting in less accumulation of starch (poor conversion from sugar to starch). However, [62] reported increased starch content in potato under heat stress. Under heat stress, PUFA content of soybean declines [63]. Vitamin C, minerals etc. are also low in seed under heat stress.

Cold stress causes stunting in plant growth, vascular browning and abnormal seed ripening. Low temperature during the flowering and seed formation phases results in pollen sterility, limiting pollen grain germination. Cold period in flowering is the most sensitive period in plants like rice. Cold stress shortens the seed filling stage and can lead to energy deficiency and thereby, sterility of gametophytes by hampering carbohydrate metabolisms. It causes cell dehydration and crystallization of water in cell. Further, there is production of ROS under low temperature (oxidative stress). Infestations of soil-borne diseases to seedling stage of the crop are also visible under cold stress. As a consequence, yield attributes and production of seeds are greatly hampered. Accumulation of minerals, amino acids, protein, starch, fat and crude fibers decreases and sugar concentration increases in seed under cold stress [31].

4.4 Heavy metal toxicity

Excessive uptake of heavy metals leads to chlorosis and reduction of production of seed through alteration of physiological, morphological, biochemical, molecular activities of plant. It negatively impacts on seed germination, accumulation and remobilization of seed reserves and photosynthetic activity resulting in loss of seed yield. Heavy metal toxicity reduces the yield also through synthesis of ROS, increased rate of lipid peroxidation and disrupting redox equilibrium. Lower uptake of nutrients such as N reduces quality *viz.*, protein of seed produce. Cd, Pb. Cr, Ni, Zn, Cu, As etc. reduce starch, protein, oil and minerals contents of seed. Decrease of oleic acid and linoleic acid and increase of palmitic acid, linolenic acid and stearic acid are found under heavy metal toxicity specially, under high Cd and Hg.

4.5 Other abiotic stresses

Use of excessive chemicals and nutrients lead to deterioration soil and water health, loss of beneficial micro-organisms activity and drastically hampers seed production and quality of the crop. On a contrary, nutrient deficiency can lead to poor yield and quality of the crop as photosynthesis and various metabolic activities are seriously affected under scarcity of nutrients. High wind causes erosion, loss of nutrients, lodging and evapotranspiration water loss which affect the plant growth and various physiological functioning inside the plant including photosynthesis. Further, high wind velocity causes loss of pollen which in turn, impacts on fertilization and seed production. Shattering loss of seeds is another major issue associated with high wind velocity. Light impacts on photosynthesis. Shading or low light hampers photosynthesis and translocation of photo-assimilates to reproductive parts and thereby, affects seed production and quality. Excessive light, on the other hand, disrupts photosynthetic apparatus including chlorophyll and hampers the photosynthetic activity of plant. Further, it creates water scarcity or high temperature, which in turn hampers various functioning of plant system and thereby, affects seed yield quantitatively and qualitatively.

5. Defensive mechanisms of plants against abiotic stresses

In response to these abiotic stresses, plant shows defensive mechanisms mainly by three ways: stress escape, stress avoidance and stress tolerance. Short growth period, seed dormancy, shedding of leaves etc. are some common escape mechanisms against abiotic stresses. Leaf rolling, reduced growth, reallocation of nutrients etc. are some common abiotic stresses avoidance strategies. In case of stress tolerance, certain cellular, physiological, biochemical and molecular changes such as osmotic adjustment, stiff cell wall, activation of stress tolerance compounds, metabolites and enzymes, expression of stress tolerance genes etc. occur inside the plant. For example, when photosynthesis is restricted, plant utilizes starch as energy source and in response to starch breakdown, synthesis of sugars, osmoprotectants, compatible solutes etc. occurs which helps the plant to tolerate stress [64]. Over expression of glutamine synthase, asparagine synthase genes etc. leads to tolerance of abiotic stresses by the plant. Besides, there are some detoxifying genes which alleviate various abiotic stresses through activating enzymes such as ascorbate peroxidase, glutathione peroxidase, glutathione reductase etc. Under abiotic stresses, abscisic acid (ABA) is produced inside the plants and this phytohormone helps the plant to cope up with abiotic stresses like drought, salinity, low temperature etc. Polyamines (low molecular weight aliphatic nitrogen compounds) also play important role in alleviating abiotic stresses through maintaining cell membrane integrity, reducing growth inhibitors, moderately expressing stress responsive genes and elevating antioxidant enzymatic activities. Endogenous polyamines can be increased by applying exogenous polyamines (putrescine, spermidine and spermine). Trace elements like selenium, silicon etc. and signaling molecules like nitric oxide also play positive roles in defending abiotic stresses.

5.1 Water stress

Drought: Plant reduces leaf area and number or modifies leaf into spine or other forms and leaf rolling to reduce transpiration loss of water. Stomata is closed and leaf abscission and curling are seen under drought stress. Cuticle or wax deposition over leaf surface also restricts loss of water through transpiration. When water is limited in soil, plant's root is extended deeper into the soil in search of moisture, but shoot growth becomes limited (through reduction of growth promoters). Plant also mature comparatively earlier than normal when exposed to drought stress. Secondary metabolites are produced under drought stress, which further improve immunity of plant. Under drought condition, synthesis of ABA in roots and its transport to shoots through xylem regulates the stomata and thereby, controls transpiration loss of water. Further, reallocation of nutrients present in older leaves occurs. Accumulation of proline (osmoprotectant), arginine (compatible solute) etc. under drought (and/or salinity stress) controls osmotic pressure and thus, helps in stress tolerance. Late embryogenesis abundant proteins are abundantly synthesized under drought stress during early embryogenesis to combat stress through improving water binding ability.

Flood: In plant like rice, adaption mechanism against flood mainly includes formation of gas filled spaces (aerenchyma) which can transfer oxygen from aerial parts to root zone of the plant.

5.2 Salinity

Accumulation ABA under high salinity regulates stomatal opening and plays key role in maintaining leaf water potential. Further, there is a production of low molecular weight organic compounds (polyols: sorbitol, mannitol, glycerol, inositol and other forms of mono and dimethylated inositol), amino acids (proline, glutamate) and betaine (betaine glycine and alanine) inside the plant as defensive mechanism to alleviate salinity stress [65]. Plant detoxifies oxidative stress enzymatically or non-enzymatically. In enzymatic system, super oxide dismutase (SOD), catalase (CAT), peroxidase (guaiacol peroxidase, glutathione peroxidase and ascorbate peroxidase), monodehydroascorbate reductase (MDHAR), glutathione reductase (GR), glutathione transferase (GST), dehydroascorbate reductase (DHAR) etc. are produced which act as ROS scavengers. In non-enzymatic system, proline, glutathione, ascorbic acid, carotenoids, flavonoids, tocopherols etc. are synthesized which neutralize ROS.

5.3 Temperature stress

Heat stress: Plants adapt heat stress by changing leaf orientation, early maturation, rolling of leaf, transpiration cooling and synthesis of antioxidants and/or osmoprotectants or stress proteins. Elevation in transcription of genes (heat shock proteins) under heat stress is one of the common plant responses against heat stress.

Cold stress: Plant adapts cold stress through acclimation. Accumulation of ABA under cold stress plays key role in plant's adaptation to it. Plant enhances the level of unsaturated lipids which in turn improve fluidity and membrane stability and thereby, defend cold stress.

5.4 Heavy metal toxicity

Plant secretes various organic acids (oxalic acid, citric acid, tartaric acid, malic acid, succinic acid) under heavy metal toxicity and these acids chelates with heavy metals resulting in conversion of toxic to non-toxic elements. Metallothionein, Phytochelatins, glutathione etc. are synthesized by the plant and they act as chelates against heavy metals. Plant root also release amino acids which provide nutrition to fungi, bacteria and others. These microorganisms inhibit the uptake of heavy metals by the plant.

5.5 Other abiotic stresses

Under heavy wind velocity, stomata are closed by the plant to reduce transpiration loss of the water. Under excess light, plant undergoes photoinhibition (downregulation of photosynthesis and activating photo-protective mechanisms to prevent light entry to chloroplast). Non-photochemical quenching and enzymatic antioxidant defense against light induced ROS are some mechanisms occur inside plant against excess light. Under shaded condition, adaptation of plant is resulted from adjusting the canopy architecture. For example, height is increased instead of branching to harvest solar energy.

6. Agronomic management practices to cope up with abiotic stresses

In order to cope up with abiotic stresses, other than the plant's own defensive mechanisms, changes in management practices play a considerable role. Few such agronomic management practices have been listed in **Table 2**. Improved breeding

Abiotic stresses	Agronomic management practices
Drought	• Use of resistant/tolerant variety • Foliar spray of 2% DAP +1% KCl or 0.5% zinc sulfate +0.3% boric acid +0.5% ferrous sulfate +1% urea or 3% kaolin or 500 ppm cycocel or 40 ppm NAA during moisture sensitive periods • Mulching or cover cropping or inter/mixed cropping to reduce evaporation loss of water (moisture conservation) • Sowing in ridge and furrow bed • Alternate/skip furrow irrigation or partial root drying • Skip row planting • Use of sprinkler/drip or any other micro irrigation/water saving options • Less application of fertilizers • Split application of N and K fertilizers • Application of Zn, B and Mn fertilizers to improve plant's tolerance against water deficit • Use of bio-fertilizers and seed priming • Nipping or pinching apical portion to arrest shoot growth and consequently, transpiration rate.
Flood	• Use of resistant variety • Foliar spray of 2% DAP +1% KCl or 0.5% zinc sulfate +0.3% boric acid +0.5% ferrous sulfate +1% urea or 3% kaolin or 500 ppm cycocel or 0.5 ppm brassinolide or 100 ppm salicylic acid • Drainage of excess water • Growing of water loving crops • Adequate application of K fertilizer

Abiotic stresses	Agronomic management practices
Salinity	• Use of resistant/tolerant variety • Foliar spray of 2% DAP +1% KCl or 0.5 ppm brassinolide or 100 ppm salicylic acid • Application of gypsum or incorporation of green manure crop in soil before sowing • Split application of N and K fertilizers • Excess (25% more) N application • Seed hardening with NaCl • Exogenous applications of ABA and/or jasmonic acid • High application of K, Ca, Mg etc. fertilizers • Seed treatment with polyamines *viz*., putrescine, spermidine, spermine etc.
High temperature	• Use of resistant/tolerant variety • Shading on the plant canopy • Use of mulch or residue retention to avoid heat stress at early growth stages • Application of salicylic acid or glycine betaine or ethylene or gibberellic acid • Moderate application of N, P, K and Ca fertilizers as they act as osmoprotectants and improve seed quality if applied at anthesis. • Irrigation on the canopy to restrict sun scorching • Drip irrigation to reduce soil temperature at root zone depth • Timely sowing of winter crops to avoid heat stress during anthesis and seed formation phases
Low temperature	• Use of resistant/tolerant variety • Seed treatment with gibberellic acid or proline • Foliar spray of 0.15% ammonium molybdate • Use of cryoprotectants, ABA, paclobutrazol, uniconazole etc. • Timely sowing of monsoon crops to avoid terminal cold stress
Heavy metal toxicity	• Construction of wetlands • Reduction of chemical based intensive farming approach • Substitution of chemicals with biofertilizer, compost and bio-pesticides
Wind velocity	• Use of windbreaks/shelterbelts • Use of dwarf crop varieties
Low light	• Use of sun loving or tall varieties
High light	• Use of shade loving or dwarf varieties
Excess chemicals and nutrients	• Promotion of organic farming practices • Less use of chemicals and nutrients • Growth of nutrient exhaustive crops
Nutrient scarcity	• Application of nutrients to correct the deficiency

Table 2.
Agronomic management practices to alleviate abiotic stresses.

program to develop highly resistant varieties is also needed. In this regard, identification of responsive genes against abiotic stresses is one of the frontline strategies that can be made by the plant breeders. Use of genetic engineering to develop transgenic plants keeping in mind the safety of the environment and human health is the another key solution against abiotic stresses. It has been also observed that use of beneficial microbes in the form of seed bio priming has helped the plant to alleviate against such abiotic stresses through its positive role in germination and early plant stand establishment [66]. Bio priming increases the osmolyte concentrations which results in high cell wall elasticity and turgid weight to dry weight ratio. Further, endophytic synthesis of alkaloids can save the macromolecules from ROS through ROS scavenging activities. Plant growth promoting rhizobacteria (PGPR) enables expression of drought response related genes through enhancing ROS scavenging activities. Synthesis of phytohormones like IAA, GA3 etc. occurs due to PGPR which in turn, helps in plant growth under stress condition. PGPR further synthesizes exopolysaccharides which improves soil structure and maintains water and nutrient uptakes. Exogenous application of various chemicals viz. proline, glycine betaine, trehalose etc., plant components such as amino acid, sugars etc. and phytohormones such as ABA, GA3, jasmonic acid, salicylic acid, brassinosteroids etc. can help plant to cope up with abiotic stresses. Salicylic acid is a phenolic phytohormone which at low level can alleviate abiotic stresses through improving stomatal regulation, leaf chlorophyll content, water use efficiency and root growth. Brassinosteroids induce expression of antioxidant genes and thus, alleviate oxidative stresses through synthesizing ABA, proline, glutathione, phytochelatins, heat shock proteins and stimulating N metabolism. Brassinosteroids and salicylic acid help in improving seed production and quality under salinity stress through osmoregulation by increase of SOD, POD, CAT activities and elevation of photosynthesis. Jasmonic acid is a cyclopentanone derivative synthesized from linolenic acid, which helps the plants to tolerate abiotic stresses. Application of silicon also plays important role in improving crop growth and productivity under abiotic stresses. Under stress conditions, plant itself synthesizes proline, L-tryptophan, glutathione (GSH), citric acid, polyols, lipoic acid, ascorbic acid, glycine betaine, α-tocopherol, melatonin etc. as defensive mechanisms. Further, exogenous application of these can be found useful in alleviating stress. Proline acts as a ROS quencher by increasing the activities of SOD, CAT etc. and maintains plant growth under drought or salinity stress. L-tryptophan (amino acid) can synthesize auxin and thus, helps in plant growth. Glutathione (GSH) is a low molecular weight tripeptide, composed of glutamine, cysteine and glycine. Application of GSH detoxifies ROS, methylglyoxal and synthesize phytochelatins which bind heavy metal. It also acts as cysteine reservoir [67]. Citric acid is the intermediate product of TCA cycle, which is produced by citrate synthase from oxaloacetate and acetyl coA. Application of citric acid shows antioxidant properties which inactivates heavy metals such as Cu, Pb, Al etc. as well as protects the plant from salinity stress. Foliar application of polyols or sugar alcohols (mannitol, sorbitol, inositol) plays positive role in osmotic adjustment through improving SOD, POD, CAT activities and thus, helps the plant to cope up with drought, salinity and heavy metal stresses. Lipoic acid application on plant canopy under salt or drought stresses reduces lipid peroxidation and increases cysteine, POD and CAT activities. Ascorbic acid or vitamin C application can neutralize ROS under drought stress. Under low temperature, seedling subjected to incubation in ascorbic acid can reduce oxidative damage and improves proline, nutrients and CAT activity. Glycine betaine application under salinity or drought stress helps the plant in osmotic adjustment and stabilization of PS II. α-Tocopherol is generally present in chloroplast and improves photosynthetic membrane integrity. Exogenous application of α-tocopherol reduces oxidative damage by scavenging ROS under abiotic stresses. Melatonin is a tryptophan derivative which increases proline synthesis and further, acts as ROS scavenger. Further, soil and foliar applications of humic substances, beneficial fungi, bacteria, chitosan, sea weed extracts etc. can play positive roles in combating abiotic stresses.

7. Conclusion and future prospects

The concept of abiotic stresses is not new. Over the years, abiotic stresses are affecting the crop with more prominent effect found in recent times as a consequence of climate change arising due anthropogenic activities. The extent of impacts from these stresses on crop varies from mild to severe, resulting in changes in seed production and quality of the produce accordingly. Under abiotic stresses, crop also shows internal defense against them by undergoing various morphological, cellular, physiological, biochemical and molecular changes, which altogether exert antagonistic impact on crop growth and yield. As food demand is increasing from enormous population growth, crop yield loss due to abiotic stresses should be addressed through incorporation of suitable modern agronomic management as well as breeding approaches. Development of resistant varieties to cope up with abiotic stresses needs special emphasis. Further, molecular research should be carried out at genetic level to study and develop suitable defensive mechanisms against such stresses. For instance, genetic engineering and genome editing are showing the prospects for resistance/ tolerance of plants against these stresses in future through transferring specific gene carrying targeted traits in crop. In achieving success in genetic engineering, identification and isolation of the key genes play fundamental role which urges for effective works of breeders. Further, it has been seen that although there are available mitigation and adaptation strategies against such stresses, they are currently insufficient. Therefore, there is an urgent need for multi-locational research trials, transfusion of modern practices, policy interventions and advances in genomics approaches to address negative impacts of stress as well as to achieve successful crop production and it can be achieved by integrating multiple approaches together rather than relying on single strategy. Participation of public and/or private organizations is highly needed in curtailing the adverse impact of abiotic stresses on crop through financial and infrastructural support, generation of modern agronomic, breeding and other relevant technologies as well as strengthening of extension service.

Author details

Rupa Das[1*] and Saikat Biswas[2]

1 Department of Seed Science and Technology, Bidhan Chandra Krishi Viswavidyalaya, Mohanpur, Nadia, West Bengal, India

2 Department of Agronomy, Bidhan Chandra Krishi Viswavidyalaya, Mohanpur, Nadia, West Bengal, India

*Address all correspondence to: rupadasbiswas18@gmail.com

References

[1] Verma S, Nizam S, Verma PK. Biotic and abiotic stress signalling in plants. Stress Signaling in Plants: Genomics and Proteomics Perspective. 2013;**1**:25-49

[2] Chaves MM, Maroco JP, Pereira JS. Understanding plant responses to drought- from genes to the whole plant. Functional Plant Biology. 2003;**30**:239-264

[3] Rao IM, Beebe SE, Polania J, Ricaurte J, Cajiao C, Garc'ıa R, Rivera M. Can tepary bean be a model for improvement of drought resistance in common bean? African Crop Science Journal. 2013;**21**:265-281

[4] Otegui ME, Slafer GA. Increasing cereal yield potential by modifying developmental traits. In: Proceedings of the 4th International Crop Science Congress, Brisbane. 2004. pp. 1-11. Available online: www.cropscience.org.au

[5] Prasad PVV, Pisipati SR, Momcilovic I, Ristic Z. Independent and combined effects of high temperature and drought stress during grain filling on plant yield and chloroplast protein synthesis elongation factor (EFTu) expression in spring wheat. Journal of Agronomy and Crop Science. 2011;**197**:430-441

[6] Acquadh. Principle of Plant Genetics and Breeding. Oxford: Willey Blackwell; 2007

[7] Kumar S. Abiotic stresses and their effects on plant growth, yield and nutritional quality of agricultural produce. International Journal of Food Science and Agriculture. 2020;**4**(4):367-378

[8] Fahad S, Bajwa AA, Nazir U, Anjum SA, Farooq A, Zohaib A, et al. Crop production under drought and heat stress: Plant responses and management options. Frontiers in Plant Science. 2017;**29**:8

[9] Bockheim JG, Gennadiyev AN. The role of soil-forming processes in the definition of taxa in soil taxonomy and the world soil reference base. Geoderma. 2000;**95**(1-2):53-72

[10] Abrol IP. Salt-affected soils: An overview. In: Chopra VL, Paroda SL, editors. Approches for Incorporating Drought and Salinity Resistance in Crop Plants. New Delhi, India: Oxford and IBH Publishing Company; 1986. pp. 1-23

[11] Levitt J. Responses of plants to environmental stresses. Academic Press. 1980;**1**:496

[12] Wuana RA, Okieimen FE. Heavy metals in contaminated soils: A review of sources, chemistry, risks and best available strategies for remediation. ISRN Ecology. 2011;**2011**:1-20. DOI: 10.5402/2011/402647

[13] Samarah NH. Effects of drought stress on growth and yield of barley. Agronomy for Sustainable Development. 2005;**25**(1):145-149

[14] Alfonso SU, Brüggemann W. Photosynthetic responses of a C3 and three C4 species of the genus Panicum with different metabolic subtypes to drought stress. Photosynthesis Research. 2012;**112**:175-191

[15] Gomez JM, Jimenz A, Olmas E, Sevilla F. Location and effects of long-term NaCl stress on superoxide dismutase and ascorbate peroxidase isoenzymes of pea (Pisum sativum cv. Puget) chloroplasts. Journal of Experimental Botany. 2004;**55**:119-130

[16] Shannon MC, Grieve CM. Tolerance of vegetable crops to salinity. Scientia Horticulturae. 1999;**78**:5-38

[17] Vu JCV, Gesch RW, Pennanen AH, Allen Hartwell L, Boote KJ, Bowes G. Soybean photosynthesis, rubisco, and carbohydrate enzymes function at supraoptimal temperatures in elevated CO_2. Journal of Plant Physiology. 2001;**158**(3):295-307

[18] Heino P, Palva ET. Signal transduction in plant cold acclimation. In: Hirt H, Shinozaki K, editors. Plant Responses to Abiotic Stress. Topics in Current Genetics. Springer Nature, Switzerland. 2003;**4**:151-186

[19] Salvucci ME, Crafts-Brandner SJ. Relationship between the heat tolerance of photosynthesis and the thermal stability of rubisco activase in plants from contrasting thermal environments. Plant Physiology. 2004;**134**:1460-1470

[20] Chen M, Thelen JJ. ACYL-LIPID DESATURASE2 is required for chilling and freezing tolerance in arabidopsis. The Plant Cell. 2013;**25**(4):1430-1444

[21] Takahashi D, Li B, Nakayama T, Kawamura Y, Uemura M. Plant plasma membrane proteomics for improving cold tolerance. Frontiers in Plant Science. 2013;**4**:90

[22] Polle A, Schützendübel A. Heavy metal signalling in plants: Linking cellular and organismic responses. In: Hirt H, Shinozaki K, editors. Plant Responses to Abiotic Stress. Topics in Current Genetics. Springer Nature, Switzerland. 2003;**4**:187-215

[23] Jiao S, Hilaire E, Guikema JA. Identification and differential accumulation of two isoforms of the CF1-b subunit under high light stress in Brassica rapa. Plant Physiology and Biochemistry. 2004;**42**:883-890

[24] Baroowa B, Gogoi N. Biochemical changes in black gram and green gram genotypes after imposition of drought stress. Journal of Food Legumes. 2014;**27**:350-353

[25] Bhandari K, Sharma KD, Hanumantha Rao B, Siddique KHM, Gaur P, Agrawal SK, et al. Temperature sensitivity of food legumes: A physiological insight. Acta Physiologiae Plantarum. 2017;**39**:68

[26] Maleki A, Naderi A, Naseri R, Fathi A, Bahamin S, Maleki R. Physiological performance of soybean cultivars under drought stress. Bulletin of Environment, Pharmacology and Life Sciences. 2013;**2**:38-44

[27] Mukamuhirwa A, Persson Hovmalm H, Bolinsson H, Ortiz R, Nyamangyoku O, Johansson E. Concurrent drought and temperature stress in rice—A possible result of the predicted climate change: Effects on yield attributes, eating characteristics, and health promoting compounds. International Journal of Environmental Research and Public Health. 2019;**16**(6):1043

[28] Wei Y, Jin J, Jiang S, Ning S, Liu L. Quantitative response of soybean development and yield to drought stress during different growth stages in the Huaibei plain, China. Agronomy. 2018;**8**:97

[29] Khan R, Srivastava R, Abdin MZ, Manzoor NM. Effect of soil contamination with heavy metals on soybean seed oil quality. European Food Research and Technology. 2013;**236**:707-714

[30] Allahmoradi P, Mansourifar C, Saiedi M, Jalali HS. Effect of different water deficiency levels on some antioxidants at different growth stages

of lentil (Lens culinaris L.). Advances in Environmental Biology. 2013;7:535-543

[31] Yagoob H, Yagoob M. The effects of water deficit stress on protein yield of mung bean genotypes. Peak Journal of Agriculture and Science. 2014;**2**:30-35

[32] Farooq M, Hussain M, Usman M, Farooq S, Alghamdi SS, Siddique KHM. Impact of abiotic stresses on grain composition and quality in food legumes. Journal of Agricultural and Food Chemistry. 2018;**66**:8887-8897

[33] Wardlaw IF, Willenbrink J. Mobilization of fructan reserves and changes in enzyme activities in wheat stems correlate with water stress during kernel filling. New Phytologists. 2000;**148**:413-422

[34] Pettigrew WT. Moisture deficit effects on cotton lint yield, yield components, and boll distribution. Agronomy Journal. 2004;**96**:377-383

[35] Torabian S, Farhangi-Abriz S, Rathjen J. Biochar and lignite affect H+-ATPase and H+-PPase activities in root tonoplast and nutrient contents of mungbean under salt stress. Plant Physiology and Biochemistry. 2018;**129**:141-149

[36] Cattivelli L et al. Drought tolerance improvement in crop plants: An integrated view from breeding to genomics. Field Crops Research. 2008;**105**:1-14

[37] Zhou R, Hyldgaard B, Yu X, Rosenqvist E, Ugarte RM, Yu S, et al. Phenotyping of faba beans (Vicia faba L.) under cold and heat stresses using chlorophyll fluorescence. Euphytica. 2018;**214**:1-13

[38] Mujeeb-Kazi A, Munns R, Rasheed A, Ogbonnaya FC, Ali N, Hollington P, et al. Breeding strategies for structuring salinity tolerance in wheat. Advances in Agronomy. 2019;**155**:121-187

[39] Bellaloui N, Mengistu A, Kassem MA. Effects of genetics and environment on fatty acid stability in soybean seed. Food and Nutrition Sciences. 2013;**4**:165-175

[40] Kazai P, Noulas C, Khah E, Vlachostergios D. Yield and seed quality parameters of common bean cultivars grown under water and heat stress field conditions. AIMS Agriculture and Food. 2019;**4**:285-302

[41] Farooq M, Nadeem F, Gogoi N, Ullah A, Alghamdi SS, Nayyar H, et al. Heat stress in grain legumes during reproductive and grain-filling phases. Crop & Pasture Science. 2017;**68**:985-1005

[42] Kang JS, Singh H, Singh G, Kang H, Kalra VP, Kaur J. Abiotic stress and its amelioration in cereals and pulses: A review. International Journal of Current Microbiology and Applied Sciences. 2017;**6**(3):1019-1045

[43] Dong B, Zheng X, Liu H, Able JA, Yang H, Zhao H, et al. Effects of drought stress on pollen sterility, grain yield, abscisic acid and protective enzymes in two winter wheat cultivars. Frontiers in Plant Science. 2017;**8**(June):1-14

[44] Ni Z, Li H, Zhao Y, Peng H, Hu Z, Xin M, et al. Genetic improvement of heat tolerance in wheat: Recent progress in understanding the underlying molecular mechanisms. Crop Journal. 2018;**6**(1):32-41

[45] Asseng S, Foster I, Turner NC. The impact of temperature variability on wheat yields. Global Change Biology. 2011;**17**(2):997-1012

[46] Jagadish S, Craufurd P, Wheeler T. High temperature stress and spikelet

fertility in rice (Oryza sativa L.). Journal of Experimental Botany. 2007;**58**(7):1627-1635

[47] Redden RJ, Vara HJL, Prasad PV, Ebert AW, Yadav SS, O'Leary GJ. Temperature, climate change, and global food security. Temperature and Plant Development. 2014;**8**:181-202

[48] Vafa P et al. The effect of drought stress on grain yield, yield components and protein content of durum wheat cultivars in Ilam Province, Iran. International Journal of Agricultural and Biosystems Engineering. 2014;**8**:631-636

[49] Triboï E et al. Environmentally-induced changes in protein composition in developing grains of wheat are related to changes in total protein content. Journal of Experimental Botany. 2003;**54**:1731-1742

[50] Flagella Z et al. Changes in seed yield and oil fatty acid composition of high oleic sunflower (Helianthus annuus L.) hybrids in relation to the sowing date and the water regime. European Journal of Agronomy. 2002;**17**:221-230

[51] Bethke PC et al. Tuber water and pressure potentials decrease and sucrose contents increase in response to moderate drought and heat stress. American Journal of Potato Research. 2009;**86**:519

[52] Singh S et al. Effect of water stress at different stages of grain development on the characteristics of starch and protein of different wheat varieties. Food Chemistry. 2008;**108**:130-139

[53] Di Caterina R et al. Influence of salt stress on seed yield and oil quality of two sunflower hybrids. Annual Applied Biology. 2007;**151**:145-154

[54] Liu Y, Li J, Zhu Y, Jones A, Rose RJ, Song Y. Heat stress in legume seed setting: Effects, causes, and future prospects. Frontiers in Plant Science. 2019;**10**:938

[55] Vaz Patto MC, Amarowicz R, Aryee ANA, Boye JI, Chung HJ, Martín-Cabrejas MA, et al. Achievements and challenges in improving the nutritional quality of food legumes. Critical Reviews in Plant Sciences. 2015;**34**:105-143

[56] Sekhon HS et al. Water use efficiency under stress environments. In: Climate Change and Management of Cool Season Grain Legume Crops. Netherlands: Springer; 2010. pp. 207-227

[57] Wilhelm I. Crop physiology and metabolism. Crop Science. 1999;**39**:1733-1741

[58] Wardlaw IF et al. Contrasting effects of chronic heat stress and heat shock on kernel weight and flour quality in wheat. Functional Plant Biology. 2002;**29**:25-34

[59] Hoegy P et al. Impacts of temperature increase and change in precipitation pattern on crop yield and yield quality of barley. Food Chemistry. 2013;**136**:1470-1477

[60] Lin CJ et al. Influence of high temperature during grain filling on the accumulation of storage proteins and grain quality in rice (Oryza sativa L.). Food Chemistry. 2010;**58**:10545-10552

[61] Sadeghipour O. The influence of water stress on biomass and harvest index in three mung bean (Vigna radiata L. (Wilczek)) cultivars. Asian Journal of Plant Sciences. 2009;**8**:245-249

[62] Silva JAB et al. Microtuberization of Andean potato species (solanum spp.) as affected by salinity. Scientia Horticulturae. 2001;**89**:91-101

[63] Dornbos DL, Mullen RE. Soybean seed protein and oil contents and fatty acid composition adjustments by drought and temperature. Journal of the American Oil Chemists' Society. 1992;**69**:228-231

[64] Krasensky J, Jonak C. Drought, salt, and temperature stress induced metabolic rearrangements and regulatory networks. Journal of Experimental Botany. 2012;**63**(4):1593-1608

[65] Grattan SR, Grieve CM. Salinity-mineral nutrient relations in horticultural crops. Scientia Horticulturae. 1998;**78**(1-4):127-157

[66] Glick BR, Cheng Z, Czarny J, Duan J. Promotion of plant growth by ACC deaminase-producing soil bacteria. European Journal of Plant Pathology. 2007;**119**:329-339

[67] Hasanuzzaman M, Nahar K, Anee TI, Fujita M. Glutathione in plants: Biosynthesis and physiological role in environmental stress tolerance. Physiology and Molecular Biology of Plants. 2017;**23**:249-268

Chapter 7

Water Stress, Heat, and Salinity in the Physiological Quality of the Seeds

Rember Pinedo-Taco, Cecilia Figueroa-Serrudo and Leonel Alvarado-Huamán

Abstract

Plant seeds, being sessile, are simultaneously exposed to favorable or adverse conditions from sowing to harvest. The physiological quality of the seed is affected by the type of biotic and abiotic stress to which the mother plant is exposed, especially in the stages of embryogenesis, development and seed filling. Therefore, the behavior of their progeny will be reflected when the seeds are capable of maintaining acceptable viability standards with a high-germination potential to generate a normal seedling and establish themselves without difficulties under field conditions. Most of the species cultivated under abiotic stress conditions reduce their physiological quality; however, some species are salt dependent, and prolonged absence of NaCl in the soil inhibits seed development, results in lower seed quality and thus limits progeny-seedling growth as is the case of Suaeda salsa, and typical annual extreme halophytic herb with succulent leaves develops well and produces high-quality seeds when grown under high salinity conditions. Consequently, the response of the plant to adverse factors depends on the genotype and its stage of development at the time of stress, the duration and severity of the type of stress and the environmental factors that cause it. Depending on the severity and duration of the stress, plants could activate mechanisms to adapt or tolerate abiotic stress conditions at the molecular, morphological, physiological and cellular levels.

Keywords: abiotic factors, climate change, heat stress, seed germination, soil salinity, water stress

1. Introduction

High-quality seed is the main input to obtain high crop yields, by producing healthy plants, resistant to diseases and to adverse conditions [1, 2]. The physical, physiological, genetic, and sanitary quality of the seeds depend on the genetic material used in sowing, management of the mother plant, climatic conditions, such as temperature, humidity, and solar intensity; likewise, the edaphic characteristics or the fertility of the soils, presence or absence of stress during the stage of germination, emergence, growth, reproduction (flowering and pollinization/fertilization, seed filling period), and harvest and post-harvest management [3–8].

Different environmental stress during seed formation that affects the final seed quality can be due to water, mineral deficiencies, salinity, and extreme temperatures among the most important; water deficiencies during grain filling, flowering, or pod formation can reduce germination potential and seed vigor [9–11]. The seed's quality is also affected when the water deficit is complemented by high temperatures, causing the production of a high proportion of small seeds with low germination and vigor [6, 12, 13]. Internal plant temperature increase causes stomatal closure and causes damage to the seed embryo due to the decrease in the thermoregulatory effect of water, thus influencing seed germination. Photosynthesis is reduced due to high temperatures and reduces the flow of the substances produced toward the seed; in this condition they are used to sustain respiration, creating an imbalance in the stage of spike formation or seed filling, and affecting the quality and size of the seeds.

On the other hand, high salinity levels generally cause a reduction in seed germination or may retard the germination process or affect plant growth by interfering with seed germination, as well as enzyme activity and unbalance mitosis mainly in glycophytic plants [14, 15]. Some species of halophyte-type plants can tolerate high levels of salinity, but the levels of germination and vigor of the seeds can be affected in the absence of salts.

In response to abiotic stress conditions, plants can activate various stress-sensitive systems, such as upregulation of stress-related proteins that function as molecular chaperones (heat shock, abundant late embryogenesis, and protein dehydrins), production of enzymatic antioxidants (catalase, superoxide dismutase, and ascorbate peroxidase), and accumulation of compatible solutes known as osmolytes (proline and glycine betaine) in plant cells subjected to osmotic stress [16]. In the case of extreme temperatures, plants show a broad structural and physiological plasticity that allows them to adapt to different temperatures [12, 17, 18].

2. Incidence of abiotic factors in seed quality

When plants grow under adverse environmental conditions, the oxidative/reductive state in their cells is altered and allows the processing of a type of protein and its transport to the nucleus to regulate the expression of genes related to the homeostasis of nitric oxide (NO) and the response to abscisic acid (ABA), a key plant hormone in the response to numerous abiotic stresses (solar radiation, temperatures, humidity, and salinity); it is a survival mechanism against different abiotic stresses during seed germination or subsequent plant growth. High radiation is rarely a problem, as long as water and nutrients are available. To obtain high yields, leaves grow and cover the soil surface as soon as possible after planting; if this process is delayed, solar radiation is lost as heat embodied in bare soil, evaporating soil moisture [19]. With high temperatures, crops need a greater amount of inputs (nutrients, water, and solar radiation) to be able to maintain its level of metabolism. During the filling of the seed or grains and as the temperature increases, the development accelerates more than the growth; even under optimal management conditions, yield can be reduced by up to 4% for every 1°C increase in mean temperature due to shortening the grain-filling period [20].

The damage caused by high temperatures is commonly associated with water stress; therefore, the availability of water becomes a critical factor for the survival and initial growth of plants and in the formation and development of seeds.

Consequently, to the extent that plants can transpire freely, they will also be able to cope with high temperatures; with enough water available they can withstand air temperatures of 40°C; however, if water is a limiting factor, leaves may die at that temperature.

Regarding salinity, salts can destroy the soil structure causing the expansion of clays and the dispersion of fine particles that clog the soil pores through which water and oxygen circulate [19]. The increase in salts causes very high values of osmotic pressure in the soil water, generally causes a reduction in seed germination or can slow down the germination process, interfering with the growth of most crops and other plants specialized [11, 15].

The environmental effects of seed production are complex, in that environment, the mother plant has a significant influence on seed traits including seed size, dormancy, and germination. In many species, factors such as the age of the mother plant and the position of the seed in the fruit, inflorescence, or canopy can affect seed properties, often accompanied by dimorphism of the seeds themselves or of the fruits themselves arise [6].

2.1 Water stress

2.1.1 Effect of water stress on seed germination

Crop production is highly influenced by the water level in the soil, one of the best conditions being field capacity. There are three degrees of water stress based on the relative water content (WRC), a measure that normalizes the real water content in the tissue with respect to the water potential that the tissue could have: (i) mild stress: decrease in the water potential of some bars (tenths of MPa) or WRC by 8–10% compared to well-watered plants under slight evaporative demand; (ii) moderate stress: decrease in water potential more pronounced, although less than −1.2 to −1.5 MPa or a decrease in WRC between 10 and 20%, and (iii) severe stress: when the decrease in water potential is greater than 15 bars (−1.5 MPa) or decrease in WRC greater than 20% [21].

Water deficit during grain filling, flowering, or seed coat formation reduces germination and seed vigor [10]. When water deficiency is complemented by high temperatures, the plant produces small seeds of poor physiological quality [12, 13].

Seed germination in *Citrullus lanatus* is extremely sensitive to water stress. An osmotic ψ of −430 kPa completely inhibits germination; compared to other cucurbits, the seeds of *C. lanatus* are the most sensitive [22]. However, the sensitivity of *C. lanatus* seeds to water stress is reduced when the seed coat and inner membrane are removed. Exposure to water stress causes secondary dormancy (lettuce seeds) when insufficient water potentials are present; when seeds are exposed to the environmental signal, secondary dormancy is overcome, which means that secondary dormancy is a reversible process for the seed [22].

2.1.2 Effect of water stress on the agronomic characteristics of the mother plant

Water stress causes a decrease in stomatal conductance that causes a reduction in transpiration and photosynthesis as a consequence of the decrease in total carbon (C) fixation. Water deficit also causes loss of starch reserves and nonstructural sugars that support the growth and development of the mother plant, due to the decrease in the activity of the photosynthetic enzyme galactinol synthase [23, 24]. Sucrose import is blocked and remobilization of starch reserves to the ovary walls occurs when water

Crop	Water stress	Effect on physiological seed quality	Source
Phaseolus vulgaris L.	During seed filling	Lower vigor due to increase in 46% CE	[26]
Zea mays L.	During anthesis	Inhibition of photosynthesis, blockade of sucrose transport, due to remobilization of starch stores to the ovary walls	[25]
Z. mays L.	Persistent water stress	Abortion of the ovary and lower yield due to the action of CWIN (cell wall invertase)	[25]
Oryza sativa L. and *Sorghum bicolor*	During flower induction and inflorescence development	Delay in panicle development and anther dehiscence	[27]
Gossypium spp.	During seed filling	Reduction of accumulation starch in the leaves and increase of hexose sugars	[27]
Various crops	During seed filling	Reduced seed development due to reduced photosynthesis	[28]
Various crops	During seed filling	Reduction of potential grain deposit, due to changes in the rate and duration of the grain filling stages.	[28]

Table 1.
Water stress and its effect on the physiological quality of seeds.

stress occurs 5 days before anthesis in the case of maize [25]. These water stress conditions directly affect the photosynthesis processes in the carbohydrate translocation from the leaves (source) to the seeds (sink) (**Table 1**) [27, 28].

In the seed filling stage of *Glycine max*, there is a reduction in yield due to water stress, in addition, in pre-flowering, the abortion of flowers occurs due to damage to the pistil (ovule) and not to the pollen, which reduces the number of pods and therefore seed yield; however, they are not affected in the physiological quality in terms of germination, vigor, and viability of the seeds [12, 26, 29, 30].

In the maize case (**Table 1**), the action of CWIN (cell wall invertase) and glucose concentrations lead to ovary abortion and yield loss in the face of permanent water deficit [25]. When water stress occurs during flower induction and inflorescence development, panicle development and anther dehiscence are delayed, resulting in a substantial reduction in rice and sorghum seed establishment [27].

2.2 Temperature stress

2.2.1 Effect of high-temperature stress on plant physiology

Most plant species are sensitive to temperature stress and suffer when these are low or very high with respect to the thresholds defined for each one. There are few production environments with ideal temperatures (5–25°C). In response to these environmental constraints, plants display a broad structural and physiological plasticity that allows them to adapt to different temperatures caused by geography, diurnal, and seasonal rhythms [17, 18]. The answers vary if it is a temporary or permanent stress, due to high night temperatures, daytime temperatures, and the daily average, or if there is an interaction between day and night temperatures. In general, four types of heat stress are recognized in plants: that caused by sustained high temperatures; frequent episodes of high temperatures ("heat shock"); damage due to cooling (from 0 to 10°C) or "chilling injury" in numerous fruits, foliage, and tropical flowers; and freeze damage at temperatures below 0°C, which causes ice formation in plant tissues [17].

High temperatures generate anatomical, morphological, and functional changes in plants, some similar to those produced by water stress. The opening of the stomata is a response of plants to stress due to high temperatures to cool their leaves through transpiration, while in response to drought they close their stomata to reduce water loss and under conditions of combined stress due to high temperatures and lack of water the stomata remain closed [31]. Heat waves are generally accompanied by periods of drought, accelerating soil water loss, and increasing the H_20 vapor pressure deficit in the air, decreasing stomatal conductance [32]. Meristems, fruits, seeds, and young leaves are closely associated with mature leaves and with the whole plant in general, through complex networks of source/sink relationships. When moderate stress occurs due to high temperatures, photosynthesis and the activity of sources and sinks are reduced simultaneously because the assimilates destined for the reproductive or conservation organs are destined to solve maintenance respiration and osmotic adjustment [33]. The photosynthetic apparatus is very susceptible to high-temperature stress; photosystem I, II, and carbon assimilation (Calvin Cycle) are the most affected processes. Consequently, the cumulative effects of these changes usually result in poor growth and reduced plant productivity [34, 35].

2.2.2 High temperatures in the physiological seed quality

The physiological seed quality can be altered due to a shorter time in the grain filling period caused by high temperatures. This stress type affects the transport and accumulation of reserve substances in the seed due to a greater fluidity of the lipids in the cell membranes of the seeds and a greater loss of electrolytes so that the seed produced under these conditions has less vigor by the low amount of reserve substances accumulated by the seeds [36, 37].

High temperatures affect the development of gametes and the double fertilization process and therefore the number of seeds formed. It also affects pollen production and viability, stigma receptivity, pollen germination, and pollen tube elongation. Therefore, exposing plants to high temperatures in early reproductive stage results in an irreversible loss of yield while in late agronomic stage losses can be reduced with some agronomic practices [38–40]. Likewise, high temperatures cause a greater production of superoxides (ROS) that could reduce the metabolic activity of the seed necessary for the germination process. Hydrogen peroxide plays important role in the germination process, however, high levels of H2O2 can be toxic to seeds [41, 42].

Reduction of the thermostability of the plasmatic membrane delays the activity of the Ca + 2 signaling, kinases, and heat shock factors are responsible for the decrease in germination and seed vigor due to thermal stress [43]. On the other hand, phytohormones related to the germination process, such as ABA and GA, have a negative interaction, especially in response to stress conditions [44]; exposure to high temperatures increases ABA concentration and decreases GA biosynthesis in seeds [45].

Likewise, substances from the group of polyamines, including putrescine, spermidine, and spermine, are recognized for their relationship with seed development and tolerance to high-temperature stress [46]. Glutathione has also been identified in tolerance to heat stress, in the regulation of redox signaling and defense processes in *Zea mays*, *Coleus blumei*, *Fagus sylvatica*, *Triticum aestivum*, and Vigna radiata [47–49].

Crop	High-temperature stress	Effect on the physiological seed quality	Source
Pisum sativum	During the production cycle	Reduced seed viability	[50, 51]
Brassica sp.	During the crop production cycle	Low seed quality and lower germination percentage	[41]
Medicago truncatula	During the grain filling phase	Regulation of imbibition and physical dormancy by changes in the properties of the seed coat	[52]
Lens culinaris	During the grain filling phase	Reduced grain-filling period, smaller seed size	[53]
Oryza sativa	During the grain filling phase	Reduction in germinative power, vigor index, and weight of 1000 seeds	[54, 55]
Lupinus sp.	During production cycle	Reduction in germinative power	[56]
Glycine max	From the start of flowering to harvest	Anomalies in flowering and fruit retention. Reduced seed quality.	[57] [58]
Brassica napus	During production cycle	Reduction in the number of fruits per plant and seed quality	[59, 60]
Cicer arietinum	During the grain filling phase	Reduction in fruit set and seed filling	[61]

Table 2.
Temperature stress effect on physiological quality of seeds.

High temperatures have favorable and in some cases favorable effects on the physiological quality of the seeds (**Table 2**). For example, in late pea, rice, and Lupinus sp. and in plants of the Brassica genus, stress due to high temperatures in the field affects crop phenology, fruit set, viability, germination, and the number of abnormal seedlings [41, 50, 54, 56]. In lentils (*Lens culinaris*), exposure to high temperatures reduces the grain filling period, as well as seed growth, the number of seeds per plant, seed weight, carbohydrate, protein, and mineral content; abnormalities in flowering reduce retention capacity of legumes, lower oil content in seeds in canola, and higher incidence of pathogens in seeds [53, 57–59].

In *Medicago truncatula*, the increase in temperature from 15 to 25°C favors the increase in the properties of the seed coat, regulating imbibition and physical dormancy [52]. Likewise, the stress environment with high temperatures in *Medicago* sp. is positive for faster and more uniform germination compared to seeds produced under optimal conditions [54]. In soybeans, exposure to high temperatures of 36/24°C and 42/26°C from the beginning of flowering to harvest, presents a lower accumulation of lipoxygenase, the β-conglycin, sucrose binding protein, and Bowman-Birk protease inhibitor compared to soybeans of the control under 28/22°C [62].

2.3 Salinity

Osmotic and water stress in plants generated by soil salinity is one of the main causes of the deterioration of arable land. It is estimated that there are more than 800 million hectares globally including 20% of cultivated land, with salinity being the cause of yield losses estimated at 20% worldwide in arid and semi-arid regions of the world [63–67].

The process of accumulation of K and Na salts in soils, with a predominance of Ca and Mg, can be of natural origin or the product of anthropogenic activities, mainly due to the indiscriminate use of chemical fertilizers, inadequate irrigation practices, and flooding of soils with seawater in coastal regions are the main causes of soil salinization [15, 68].

There are plants that can tolerate or be completely susceptible to salinity. Most cultivated plants are salt-sensitive glycophytes. In contrast, halophytic species such as quinoa (*Chenopodium quinoa* Willd.) and *Suaeda aralocapsica*, are capable of reducing stomatal density when grown under hypersalinity conditions [5, 11, 69]. Quinoa can tolerate high levels of salinity (6 d S m -1), without a significant decrease in seed yield and total biomass [11].

Salt stress affects seed germination rate, germination initiation, and seedling establishment due to osmotic stress, ion toxicity, and oxidative stress [11, 14]. The presence of salts in the soil can reduce seed germination by decreasing the amounts

Crop	Evaluation/salinity and stress	Effect on physiological seed quality	Source
Zea mays L.	Concentration 4, 6, 8 dS m^{-1}	Germination delay and decrease in dry matter	[70]
Helianthus annuus L.	200 mmol NaCl	Lower percentage of germination	[71]
Z. mays L.	0 to 8 mmol NaCl	Lower percentage of germination	[68]
Z. mays L.	NaCl (0 g/L, 4.24 g/L and 7.06)	Lower percentage of germination	[15]
Z. mays L.	Concentration of 8 dS m^{-1}	Lower percentage of germination	[70]
Passiflora edulis f. flavicarpa	Saline treatments from 0.75; 2.5; 4.5 and 6.5 $dS.m^{-1}$	79% emergency with (0.75 $dS.m^{-1}$) and 48.6% with 6.5 $dS.m^{-1}$	[72]
Chenopodium quinoa Willd	Germination test	Lower percentage of germination	[73]
Conyza canadensis		Lower percentage of germination	[74]
X. *Triticosecale Wittmarck* and H. vulgare	Solutions of $MgCl_2$ and $CaCl_2$, concentrations of NaCl: 2.56, 5.12, 7.68 and 10.54 g	Tolerance to solutions of extreme salinity, good germination percentage	[75]
Salicornia europaea	Hypersalinity (80 dS m^{-1})	Inhibits seed germination almost completely	[76]
Suaeda salsa	200 mM NaCl, during growth	Greater seed development, greater potential for seedling emergence	[5]
Suaeda salsa	Prolonged absence of NaCl	Inhibits seedling development and affects seed quality	[5]
S. europaea L.	Chloride salts (from 0.5–1%) and sulfate salts in 0.5–3%	Stimulating effects on germination	[77]
S. ciliata	340 mM NaCl	It inhibits germination almost completely	[78]
Glycine max	200 mM NaCl	Lower percentage of germination	[79]
Cynodon sp.	100 mM	Lower percentage of germination	[80]

Table 3.
Salinity stress effect on physiological quality of seeds.

of gibberellic acids (GA), increasing the levels of abscisic acid (ABA), and altering cell membrane permeability and water behavior within the seed [64]. However, the presence of salts may not always be detrimental to all crops (**Table 3**); some species require salts from the germination stage [5, 11]. In halophytic plant types such as *S. salsa*, the prolonged absence of NaCl inhibits development and affects seed quality; as a consequence, it limits the growth of progeny seedlings in terms of biomass and seed yield [76, 81]. In C. quinoa and *S. aralocapsica* with 0.3 M NaCl and 1.5 M NaCl, the percentage of seed germination is reduced between 10 and 20% [5, 11, 69]. In the case of rice seeds, they are relatively tolerant during germination and very sensitive at the seedling stage; after establishment, plant tolerance increases progressively until panicle differentiation, and decreases again until the flowering stage [82].

2.3.1 Salinity in the growth and development of the mother plant

The growth conditions experienced by the mother plants affect the quality and behavior of the next generation, expressed in the physiological quality, size, and weight of the seed [8, 68]. Salinity stress during germination and early seedling growth affects crop growth and yield [83]. Shoot and root length are strongly affected by increased salinity of a stressed seed [79].

Stomatal conductance, transpiration rate, and CO2 concentration in cells decrease under salt stress affecting growth, development, and crop yield [38]. Ionic toxicity causes metabolic imbalance and protein synthesis in saline soils and also limits plant growth due to the replacement of K^+ by Na^+; biochemical reactions and conformational changes induced by Na^+ and Cl^- occur in proteins [84–87]. In saline soils, the high concentration of toxic ions in the rhizosphere is a function of their level of interaction with mineral nutrients. The interaction of salts can result in considerable nutritional deficit and imbalance. The ionic imbalance in conditions of high soil salinity occurs in the cells due to the excessive accumulation of Na^+ and Cl^- ions that reduces the uptake of other mineral nutrients such as K^+, Ca^{2+}, and Mn^{2+} [88]. The adverse effects of salinity on plant development are most profound during the reproductive phase and lead to cell cycle imbalance and differentiation. In trees, salt restricts the cell cycle by interfering with cyclin and kinase activities within the plant system and thus produces fewer cells in the meristem, limiting growth [89].

Faced with this type of stress, crops have complex physiological and biochemical response mechanisms and various factors, both inherent to the genotype, and to the morphology and physiology of the plant, influence these adverse conditions [87, 90, 91]. Proline, as an important osmoprotectant, contributes to osmotic adjustment, protecting enzymes from oxidative damage in saline conditions [67, 87]. The accumulation of other compounds such as soluble sugar facilitates the maintenance of turgor and/or the protection of the macromolecular structure against the destabilizing effects of decreased water activity [92].

In case of quinoa, considered facultative halophyte, it can grow at a high level of salinity up to 18 dS m^{-1}, without having a decrease in seed yield and biomass with a salinity of up to 6 dS m^{-1} [11]. The accumulation of endogenous hormones at different NaCl concentrations during plant growth may be related to seed development and salt tolerance of brown and black Suaeda sauce seeds. These characteristics may help the species ensure seedling establishment and population succession in variable saline environments [93].

2.3.2 Salinity effect on seed germination

Seed germination is regulated by internal factors, such as proteins, plant hormones (gibberellins/ABA, ethylene, and auxin), related genes (maturation genes and genes regulating hormones and epigenetics), nonenzymatic processes, seed age, size of the seed, and structural components of the seed, including (endosperm and seed coat) and by external factors, such as soil moisture, light, salinity, temperature, acidity, and nutrients [94–96]. Plants subjected to salinization are affected from germination to more advanced stages of development; the presence of salts interferes with the water potential of the soil, reducing the potential gradient between the soil and the seed, restricting water absorption [70, 74, 97, 98]. When the osmotic potential of the solution is lower than that of the embryonic cells, the speed, the percentage of germination, and the formation of seedlings are reduced [64, 71, 72].

Changing the physiological activity of the seed can affect the protein content and therefore the nutrient reserve in the endosperm or cotyledons, affecting the processes of seed germination and with low vigor indices, as occurs in species such as broccoli and cauliflower. Absorption of excess Na^+ and Cl^- ions from soils creates ionic stress and causes toxicity that contributes to the disruption of biochemical processes, including nucleic and protein metabolism, energy production, and respiration [99].

Salinity can negatively influence germination or delay seed germination by decreasing amounts of seed germination stimulants, such as GA, increasing amounts of ABA, and altering membrane permeability and water behavior in the seed. Delay in water uptake and a decrease in α-amylase activity with an increase in NaCl concentration may be the main reasons for delayed germination time [70].

Under saline conditions, the physiological quality of the seeds in each cultivar or species can have variable behaviors (**Table 3**). The grass species ryegrass (*Lolium perenne* L.); barley (*Hordeum vulgare* L.); vetch (*Vicia sativa* L.) and *Cicer arietinum* L.; alfalfa (*Medicago sativa* L.); oats (*Avena sativa* L.) under conditions of 0, 50 and 100 mM NaCl show a germination percentage greater than 70% [80]. However, the germination rate of all species is reduced to 200 and 400 mM respectively, producing an important reduction in water absorption levels compared to seeds not subjected to salt stress [80]. In the case of barley, as it is a highly salt-tolerant crop, it can germinate at 400 mM of NaCl, reaching 24% germination with a reduction of 76% compared to the control. The reduction in the percentage of seed germination is due to the reduction in germination with the increase in NaCl concentrations, it is the result of a decrease or delay in the absorption of water in the seeds due to the toxic effects that the ions exert on them since the functions of the membrane and the cell wall of the embryo are affected; as a result of the plasmatic membranes permeability reduction, an accumulation of external ions and loss of cytosolic solutes [100].

Some species are salt dependent, prolonged absence of NaCl in the soil inhibits seed development, results in lower seed quality and thus limits progeny seedling growth as is the case of *Suaeda salsa*, typical annual extreme halophytic herb with succulent leaves, develops well and produces high-quality seeds when is grown under high salinity conditions [5].

3. Conclusion

High-quality seed is the main input to obtain high crop yields; the physical, physiological, genetic, and sanitary quality of the seeds depend on the genetic material used in sowing, management of the mother plant, temperature conditions, humidity, solar intensity, and soil fertility. The environmental effects of seed production are complex, the environment of the mother plant has a significant influence on seed traits, including seed size, dormancy, and germination. Seed germination is regulated by internal factors, such as proteins, plant hormones (gibberellins/ABA, ethylene, and auxin), related genes (maturation genes and hormone and epigenetic regulatory genes), nonenzymatic processes, seed age, size of the seed and structural changes, seed components, including (endosperm and seed coat), and by external factors, such as soil moisture, light, salinity, temperature, pH, and nutrients.

During the seed formation stage, the final quality of the seed can be affected by water, mineral, salinity, and temperature deficiencies. Water deficiency during grain filling, flowering, or pod formation reduces germination potential and seed vigor and damage can be greater when water deficit is complemented by high temperatures, causing the production of a high proportion of small seeds due to loss of nonstructural starch and sugar reserve. At the level of the mother plant, water deficit affects the growth and development of the mother plant, due to the decrease in the activity of the photosynthetic enzyme galactinol synthase.

The increase in the internal temperature of the plant causes the closure of the stomata and causes damage to the seed due to the decrease in the thermoregulatory effect of water, which influences seed germination. The decrease in photosynthesis at high temperatures reduces the flow of the substances produced toward the seed; in this condition they are used to sustain respiration, creating an imbalance in the stage of spike formation or seed filling, and affecting the quality and size of the seeds. Consequently, the cumulative effects of these changes often result in poor growth and reduced plant productivity.

High temperatures affect the development of gametes and the process of double fertilization and therefore the number of seeds formed. It also affects pollen production and viability, stigma receptivity, pollen germination, and pollen tube elongation. In high-temperature conditions, there is a higher production of superoxides (ROS) reducing the metabolic activity of the seed necessary for the germination process.

High salinity levels reduce the germination potential of seeds or may retard the germination process or affect plant growth by interfering with seed germination as well as enzyme activity and unbalance mitosis mainly in glycophytic plants. Some plant species of the halophytic type can tolerate high levels of salinity, but the levels of germination and vigor of the seeds can be affected in the absence of salts. Most cultivated plants are salt-sensitive glycophytes. In contrast, halophytic species such as quinoa (*C. quinoa* Willd.) and *Suaeda aralocapsica* are capable of reducing stomatal density when grown under hypersalinity conditions. Quinoa can tolerate and function without a significant decrease in seed yield and biomass in salinity up to 6 d S m $^{-1}$, in halophytes, the absence.

Author details

Rember Pinedo-Taco*, Cecilia Figueroa-Serrudo and Leonel Alvarado-Huamán
Universidad Nacional Agraria La Molina, Lima, Perú

*Address all correspondence to: rpinedo@lamolina.edu.pe

References

[1] Bishaw Z, Niane AA, Gan Y. Quality seed production. In: En Yadav SS, DL MN, Stevenson PC, editors. Lentil. An Ancient Crop for Modern Times. Holanda: Springer; 2007. pp. 349-383

[2] Nóbrega JS, Lopes KP, Santos JB, dos Paiva FJ, da Silva JG, da Lima GS, et al. Quality of sesame seeds produced under soil salinity levels. Pesquisa Agropecuária Tropical. 2018;**48**(3):280-286

[3] Bass LN. Seed viability during long-term storage. Horticultural Reviews. 1980;**2**:117-141

[4] Gan Y, Stobbe EH, Moes J. Relative date of wheat seedling emergence and its impact on grain yield. Crop Science. 1992;**32**:1275-1281

[5] Guo J, Du M, Tian H, Wang B. Exposure to high salinity during seed development markedly enhances seedling emergence and fitness of the progeny of the extreme Halophyte Suaeda salsa. Frontiers in Plant Science. 2020;**11**:1291. DOI: 10.3389/fpls.2020.01291

[6] Penfield S, MacGregor DR. Effects of environmental variation during seed production on seed dormancy and germination. Journal of Experimental Botany. 2017;**68**(4):819-825. DOI: 10.1093/jxb/erw436

[7] Singh RP, Prasad PVV, Reddy KR. Impacts of changing climate and climate variability on seed production and seed industry. Advances in Agronomy. 2013;**118**:49-110. DOI: 10.1016/b978-0-12-405942-9.00002-5

[8] Tanveer A, Tasneem M, Khaliq A, Javaid M, Chaudhry MN. Influence of seed size and ecological factors on the germination and emergence of field bindweed (*Convolvulus arvensis*). Planta Daninha. 2013;**31**:39-51

[9] Rocha MR, Martins CC, Martins D, Navas da Silva RJ. Estresse hídrico induzido por soluções de PEG e de NaCl na germinação de sementes de nabiça e fedegoso. Bioscience Journal. 2014;**30**(3):687-696

[10] Franca NJB, Krzyzanowski FC, Henning AA, West SH, Miranda LC. Soybean seed quality as affected by shriveling due to heat and drought stresses during filling. Seed Science and Technology. 1993;**21**:107-116

[11] Bosque H, Rodríguez JP. Physiology of Quinoa in Saline Conditions. In: Varma A, editor. Biology and Biotechnology of Quinoa. Singapore: Springer; 2021. DOI: 10.1007/978-981-16-3832-9_10

[12] Dornbos DL, Müllen RE. Influence of stress during soybean seed fill on seed weight, germination and seedling growth rate. Canadian Journal of Plant Science. 1999;**71**:373-383

[13] Lin TY, Markhart AH. Phaseolus acutifolius a gray is more heat tolerant than P. vulgaris L. in the absence of water stress. Crop Science. 1996;**36**:110-114

[14] Kibria MG, Hoque MA. A review on plant responses to soil salinity and amelioration strategies. Open Journal of Soil Science. 2019;**9**:219-231. DOI: 10.4236/ojss.2019.911013

[15] Laynez-Garsaball JA, Méndez-Natera JR, Mayz-Figueroa J. Efecto de la salinidad y del tamaño de la semilla sobre la germinación y crecimiento de plántulas de maíz (*Zea mays* l.) bajo condiciones de laboratorio. Tip Revista Especializada

en Ciencias Químico-Biológicas. 2008;**11**(1):17-25

[16] Salleh MS, Nordin MS, Puteh AB. Germination performance and biochemical changes under drought stress of primed rice seeds. Seed Science and Technology. 2020;**48**(3):333-343. DOI: 10.15258/sst.2020.48.3.02

[17] Larcher W. Physiological Plant Ecology. 2nd ed. Berlin, Alemania: Springer-Verlag; 1980

[18] Kappen L. Ecological significance of resistance to high temperature. In: Lange OL et al., editors. Physiological Plant Ecology. Responses to the Physical Environment. Vol. 12A. Berlin, GER.: Springer-Verlag; 1981. pp. 439-474

[19] Rawson H, Gómez H. Trigo Regado. Organización de las Naciones Unidas para la Agricultura y la Alimentación: Manejo del cultivo; 2001

[20] Stapper M, Fischer RA. Genotype, sowing date and plant spacing influence in high-yielding irrigated wheat in southern New South Wales. III. Potential yields and optimum flowering dates. Australian Journal of Agricultural Research. 1990;**41**:1043-1056

[21] Hsiao TC. Plant responses to water stress. Annual Review of Plant Physiology. 1973;**24**(1):519-570. DOI: 10.1146/annurev.pp.24.060173

[22] Botha FC, Grobbelaar N, Small JGC. The effect of water stress on the germination of *Citrullus lanatus* seeds. South African Journal of Botany. 1984;**3**:111-114

[23] Pattanagual P, Madore MA. Water deficit effects on raffinose family oligosaccharide metabolism in Coleus. Plant Physiology. 1999;**121**:987-993

[24] Zinselmeier C, Byeong-Ryong J, Boyer JS. Starch and the control of kernel number in maize at low water potentials. Plant Physiology. 1999;**121**:25-36

[25] Hui-Yong L, Lübberstedt T. Molecular mechanisms controlling seed set in cereal crop species under stress and non-stress conditions. Journal of Integrative Agriculture. 2018;**17**(5):965-974

[26] Castañeda Saucedo MC, Córdova Téllez L, González Hernández V, Delgado Alvarado A, Santacruz Varela A, García de los Santos G. Respuestas fisiológicas rendimiento y calidad de semilla en frijol sometido a estrés hídrico. Interciencia. 2006;**31**(6):461-466

[27] Kolahi M, Faghani E, Goldson-Barnaby A, Sohrabi B. Physiological traits and anatomic structures of the seed for two short cotton season genotypes (*Gossypium hirsutum* L.) under water stress. Journal of Integrative Agriculture. 2020;**19**(1):89-98

[28] Nawaz H, Hussain N, Ahmed N, Rehman H, Alam J. Efficiency of seed bio-priming technique for healthy mungbean productivity under terminal drought stress. Journal of Integrative Agriculture. 2021;**20**(1):87-99

[29] Kokubun M, Shimada S, Takahashi M. Flower abortion caused by preanthesis water deficit is not attributed to impairment of pollen soybean. Crop Science. 2001;**41**:1517-1521

[30] Nielsen DC, Nelson NO. Black bean sensitivity to water stress at various growth stages. Crop Science. 1998;**38**:422-427

[31] Cohen I, Zandalinas SI, Fritschi FB, Sengupta S, Fichman Y, Azad RK, et al. The impact of water deficit and heat stress combination on the molecular response, physiology, and seed production of soybean. Physiologia Plantarum. 2021;**172**(1):41-52. DOI: 10.1111/ppl.13269

[32] McDowell NG, Beerling DJ, Breshears DD, Fisher RA, Raffa KF, Stitt M. The interdependence of mechanisms underlying climate-driven vegetation mortality. Trends in Ecology & Evolution. 2011;**26**:523-532. DOI: 10.1016/j.tree.2011.06.003

[33] Taiz L, Zeiger E. Plant Physiology. 5th ed. Sunderland: Sinauer Associates Inc.; 2010. p. 782

[34] Essemine J, Govindachary S, Ammar S, Bouzid S, Carpentier R. Abolition of photosystem I cyclic electron flow in Arabidopsis thaliana following thermal stress. Plant Physiology and Biochemistry. 2011;**49**:235-243

[35] Wahid AS, Gelani M, Ahsraf M, Fooland R. Heat tolerance in plants: An overview. Environmental and Experimental Botany. 2007;**61**:199-223

[36] Porch TG, Hall AE. Heat tolerance. In: Kole C, editor. Genomics and Breeding for Climate-Resilient Crops. Vol. 2. Berlin, GER: Springer-Verlag; 2013. pp. 167-202

[37] Shinohara T, Hampton JG, Hill MJ. Effects of the field environment before and after seed physiological maturity on hollow heart occurrence in garden pea (*Pisum sativum*). New Zealand Journal of Crop and Horticultural Science. 2006;**34**:247-255. DOI: 10.1080/01140671.2006.9514414

[38] Liu Y, Li J, Zhu Y, Jones A, Rose RJ, Song Y. Heat stress in legume seed setting: Effects, causes, and future prospects. Frontiers in Plant Science. 2019;**10**:938. DOI: 10.3389/fpls.2019.00938

[39] Kumar S, Thakur P, Kaushal N, Malik JA, Gaur P, Nayyar H. Effect of varying high temperatures during reproductive growth on reproductive function, oxidative stress and seed yield in chickpea genotypes differing in heat sensitivity. Archives of Agronomy and Soil Science. 2013;**59**(6):823-843. DOI: 10.1080/03650340.2012.68342

[40] Siebers MH, Yendrek CR, Drag D. Heat waves imposed during early pod development in soybean (Glycine max) cause significant yield loss despite a rapid recovery from oxidative stress. Global Change Biology. 2015;**21**:3114-3125. DOI: 10.1111/gcb.12935

[41] Rashid M, Hampton JG, Shaw ML, Rolston MP, Khan KM, Saville DJ. Oxidative damage in forage rape (Brassica napus L.) seeds following heat stress during seed development. Journal of Agronomy and Crop Science. 2020;**206**:101-117. DOI: 10.1111/jac.12372

[42] Wojtyla Ł, Lechowska K, Kubala S, Garnczarska M. Different modes of hydrogen peroxide action during seed germination. Frontiers in Plant Science. 2016;**7**:66

[43] Fahad S, Bajwa AA, Nazir U, Anjum SA, Farooq A, Zohaib A. Crop production under drought and heat stress: Plant responses and management options. Frontiers in Plant Science. 2017;**8**:1147. DOI: 10.3389/fpls.2017.01147

[44] Begcy K, Sandhu J, Walia H. Transient heat stress during early seed development primes germination and seedling establishment in rice. Frontiers in Plant Science. 2018;**9**:1768. DOI: 10.3389/fpls.2018.01768

[45] Toh S, Imamura A, Watanabe A, Nakabayashi K, Okamoto M, Jikumaru Y. High temperature-induced abscisic acid biosynthesis and its role in the inhibition of gibberellin action in Arabidopsis seeds. Plant Physiology. 2008;**2008**(146):1368-1385. DOI: 10.1104/pp.107.113738

[46] Zhou R, Hu QJ, Pu Q, Chen MX, Zhu XR, Gao C, et al. Spermidine enhanced free polyamine levels and expression of polyamine biosynthesis enzyme gene in rice spikelets under heat tolerance before heading. Scientific Reports. 2020;**10**:8976

[47] Kocsy G, Szalai G, Galiba G. Induction of glutathione synthesis and glutathione reductase activity by abiotic stresses in maize and wheat. Scientific World Journal. 2002;**2**:1699-1705

[48] Nieto-Sotelo J, Ho T-HD. Effect of heat shock on the metabolism of glutathione in maize roots. Plant Physiology. 1986;**1986**(82):1031-1035

[49] Nahar K, Hasanuzzaman M, Alam MM, Fujita M. Exogenous glutathione confers high temperature stress tolerance in mung bean (*Vigna radiata* L.) by modulating antioxidant defense and methylglyoxal detoxification system. Environmental and Experimental Botany. 2015;**112**:44-54

[50] Lamichaney A, Parihar AK, Hazra KK, Dixit GP, Katiyar PK, Singh D, et al. Untangling the influence of heat stress on crop phenology, seed set, seed weight, and germination in field pea (*Pisum sativum* L.). Front Plant Science. 2021;**12**:635868. DOI: 10.3389/fpls.2021.635868

[51] Jiang Y, Lahlali R, Karunakaran C, Kumar S, Davis AR, Bueckert RA. Seed set, pollen morphology and pollen surface composition response to heat stress in field pea. Plant, Cell & Environment. 2015;**38**(11):2387-2397. DOI: 10.1111/pce.12589

[52] Renzi JP, Duchoslav M, Brus J, Hradilová I, Pechanec V, Václavek T. Physical dormancy release in Medicago truncatula seeds is related to environmental variations. Plants. 2020;**9**:503. DOI: 10.3390/plants9040503

[53] Sehgal A, Sita K, Bhandari K, Kumar S, Kumar J, Vara Prasad PV, et al. Influence of drought and heat stress, applied independently or in combination during seed development, on qualitative and quantitative aspects of seeds of lentil (*Lens culinaris* Medikus) genotypes, differing in drought sensitivity. Plant, Cell & Environment. 2019;**42**(1):198-211. DOI: 10.1111/pce.13328

[54] Chen Z, Ly VJ, Ly VB, Buitink J, Leprince O, Verdier J. Genome-wide association studies of seed performance traits in response to heat stress in *Medicago truncatula* uncover MIEL1 as a regulator of seed germination plasticity. Frontiers in Plant Science. 2021;**12**:673072. DOI: 10.3389/fpls.2021.673072

[55] Folsom JJ, Begcy K, Hao X, Wang D, Walia H. Rice fertilization-Independent Endosperm1 regulates seed size under heat stress by controlling early endosperm development. Plant Physiology. 2014;**165**(1):238-248. DOI: 10.1104/pp.113.232413

[56] Omoarelojie LO, Kulkarni MG, Finnie JF, Pospisil T, Strnad M, Van Staden J. Synthetic strigolactone (rac-GR24) alleviates the adverse effects of heat stress on seed germination and photosystem II function in lupine seedlings. Plant Physiology and Biochemistry. 2020;**155**:965-979. DOI: 10.1016/j.plaphy.2020.07.043

[57] Ferreira DB, Rao VB. Recent climate variability and its impacts on soybean yields in Southern Brazil. Theoretical and Applied Climatology. 2011;**105**:83-97

[58] Lima JJP, Buitink J, Lalanne D, Rossi RF, Pelletier S, Da Silva EAA, et al. Molecular characterization of the

acquisition of longevity during seed maturation in soybean. PLoS One. 2017;**12**:e0180282

[59] Elferjani R, Soolanayakanahally R. Canola responses to drought, heat, and combined stress: Shared and specific effects on carbon assimilation, seed yield, and oil composition. Frontiers in Plant Science. 2018;**9**:1224. DOI: 10.3389/fpls.2018.01224

[60] Brunel-Muguet S, D'Hooghe P, Bataille MP, Larre C, Kim TH, Trouverie J, et al. Heat stress during seed filling interferes with sulfur restriction on grain composition and seed germination in oilseed rape (Brassica napus L.). Front Plant Science. 2015;**6**:213. DOI: 10.3389/fpls.2015.00213

[61] Awasthi R, Kaushal N, Vadez V, Turner NC, Berger J, Siddique KHM, et al. Individual and combined effects of transient drought and heat stress on carbon assimilation and seed filling in chickpea. Functional Plant Biology. 2014;**41**(11):1148-1167. DOI: 10.1071/FP13340

[62] Krishnan HB, Kim WS, Oehrle NW, Smith JR, Gillman JD. Effect of heat stress on seed protein composition and ultrastructure of protein storage vacuoles in the cotyledonary parenchyma cells of soybean genotypes that are either tolerant or sensitive to elevated temperatures. International Journal of Molecular Sciences. 2020;**21**(13):4775. DOI: 10.3390/ijms21134775

[63] FAO. Mejoramiento de Maíz con objetivos especiales. Disponible en: www.fao.rog/DOCREP/003/X7650S/x7650s21, 2015. Consultado el 13/02/2019

[64] Pereira IC, Catão HCRM, Caixeta F, Pereira IC, Catão HCRM, Caixeta F. Seed physiological quality and seedling growth of pea under water and salt stress. Revista Brasileira de Engenharia Agrícola e Ambiental. 2020;**24**(2):95-100. DOI: 10.1590/1807-1929/agriambi.v24n2p95-100

[65] Gómez-Padilla E, Ruiz-Díez B, Fajardo S. Caracterización de rizobios aislados de nódulos de frijol caupí, en suelos salinos de Cuba. Cultivos Tropicales, 2017. 2017;**38**(4):39-49

[66] Jamil A, Riaz S, Ashraf M, Foolad MR. Gene expression profiling of plants under salt stress. CRC Critical Reviews in Plant Science. 2011;**30**(5):435-458

[67] Ashraf M, Harris PJC. Potential biochemical indicators of salinity tolerance in plants. Plant Science. 2004;**166**:3-16. DOI: 10.1016/j.plantsci.2003.10.024

[68] Souza-Leme E, Oliveira S, Tavares L, Mendonça A, Dias-Leitzke I, Meneghello G, et al. Productividad y calidad fisiológica de las semillas de arroz irrigadas bajo estrés salino y fertilización con ceniza de cáscara de arroz carbonizada. Agrociencia. 2016;**50**(3):307-321

[69] Uçarlı C, Fahad S, Saud S, Chen Y, Chao W, Depeng W. Effects of salinity on seed germination and early seedling stage. In: En Abiotic Stress in Plants (Ch. 11. London: IntechOpen; 2021. DOI: 10.5772/intechopen.93647

[70] Sangoquiza Caiza C, Viera Tamayo Y, Yánez Guzmán C, Zambrano Mendoza JL. Efecto del estrés salino sobre el crecimiento de plántulas de maíz variedad "Tayuyo" en condiciones in vitro. Centro Agrícola. 2021;**48**(2):14-23 Epub 01 de abril de 2021

[71] Wu QG, Jiao Q, Shui QZ. Effect of salinity on seed germination, seedling growth, and inorganic and organic

solutes accumulation in sunflower (*Helianthus annuus* L). Plant, Soil and Environment. 2015;**61**:220-226

[72] Meza N, Arizaleta M, Bautista D. Efecto de la salinidad en la germinación y emergencia de semillas de parchita (*Passiflora edulis* f. flavicarpa). Revista de la Facultad de Agronomía. 2007;**24**(1):69-80

[73] Jacobsen SE, Hill J, Stølen O. Stability of quantitative traits in quinoa (*Chenopodium quinoa*). Theoretical and Applied Genetics. 1996;**93**(1):110-116

[74] Mitsuo Yamashita O, Carneiro Guimarães S. Efecto del estrés salino sobre la germinación de las semillas de *Conyza canadensis* y *Conyza bonariensis*. Bioagro. 2011;**23**(3):169-174

[75] Choque-Marca W, Paco-Pérez V, Murga-Cruz L. Respuesta de la germinación de semillas forrajeras a soluciones salinas en condiciones controladas. Journal of the Selva Andina Biosphere. 2020;**8**(2):69-79

[76] Araus J, Zahra F, Thushar S, Shashid M, Elouafi I, Bort J, et al. Effect of irrigation salinity and ecotype on the growth, physiological indicators and seed yield and quality of *Salicornia europaea*. Plant Science. 2021;**304**. DOI: 10.1016/j.plantsci.2021.110819

[77] Orlovsky N, Japakova U, Zhang H, Volis S. Effect of salinity on seed germination, growth and ion content in dimorphic seeds of Salicornia europaea L. (Chenopodiaceae). Plant diversity. 2016;**38**(4):183-189. DOI: 10.1016/j.pld.2016.06.005

[78] Vieira CC. Effects of salinity on seed germination, seedling growth and survival of *Spartina ciliata* Brong. Acta Botânica Brasílica. 1999;**13**(3):317-322

[79] Pavli OI, Foti C, Skoufogianni G, Karastergiou G, Panagou A. Effect of salinity on seed germination and seedling development of soybean genotypes. International Journal of Environmental Sciences & Natural Resources. 2021;**27**(2):556210. DOI: 10.19080/IJESNR.2021.27.556210

[80] Lastiri MA, Álvarez D, Soria LH, Ochoa S, Cruz-Cárdenas G. Efecto de la salinidad en la germinación y emergencia de siete especies forrajeras. Revista Mexicana de Ciencias Agrícolas. 2017;**8**(6):1245-1257

[81] Madueño-Molina A, García-Paredes JD, Martínez-Hernández J, Bugarín-Montoya R, Bojórquez-Serrano JI. Salinidad inducida con NaCl y aplicación de fósforo sobre las propiedades bioquímicas de frijolillo (*Rhynchosia minima* L (DC)). Universidad y ciencia. 2011;**27**(1):43-51

[82] Yoshida S. Fundamental of Rice Crop Science. International Rice Research Institute: Los Baños, Laguna, Philippines; 1981. p. 269

[83] Kitajima K, Fenner M. Ecology of seedling regeneration. In: Fenner M, editor. Seeds: The ecology of regeneration in plant communities. 2nd ed. Wallingford, UK: CABI Publishing; 2000. pp. 331-359

[84] Assaha DVM, Ueda A, Saneoka H, Al-Yahyai R, Yaish MW. The role of Na(+) and K(+) transporters in salt stress adaptation in glycophytes. Frontiers in Physiology. 2017;**8**:509. DOI: 10.3389/fphys.2017.00509

[85] Zhu JK. Salt and drought stress signal transduction in plants. Annual Review of Plant Biology. 2002;**53**:247-273. DOI: 10.1146/annurev.arplant.53.091401.143329

[86] Kronzucker HJ, Coskun D, Schulze LM, Wong JR, Britto DT. Sodium

as nutrient and toxicant. Plant and Soil. 2013;**369**(1):1-23

[87] Gupta B, Huang B. Mechanism of salinity tolerance in plants: Physiological, biochemical, and molecular characterization. International Journal of Genomics. 2014; 701596. DOI: 10.1155/2014/701596

[88] Maathuis F. Sodium in plants: Perception, signalling, and regulation of sodium fluxes. Journal of Experimental Botany. 2014;**65**(3):849-858

[89] Shristava P, Kumar R. Soil salinity: A serious environmental issue and plant growth promoting bacteria as one of the tools for its alleviaton. Saudi Journal of Biological Sciences. 2015;**22**(2):123-131. DOI: 10.1016/j.sbjs.2014.12.001

[90] Orosco-Alcalá BE, Núñez-Palenius HG, Pérez-Moreno L, Valencia Posadas M, Trejo-Téllez LI, Díaz-Serrano FR, et al. Tolerancia a salinidad en plantas cultivadas: una visión agronómica. Agroproductividad. 2018;**11**:51-57

[91] Radić S, Štefanić PP, Lepeduš H, Roje V, Pevalek-Kozlina B. Salt tolerance of Centaurea ragusina L. is associated with efficient osmotic adjustment and increased antioxidative capacity. Environmental and Experimental Botany. 2013;**87**:39-48

[92] Florido M, Bao L. Tolerancia a estrés por déficit hídrico en tomate (*Solanum lycopersicum* L.). Cultivos Tropicales. 2014;**35**(3):70-88

[93] Wang F, Xu YG, Wang S, Shi W, Liu R, Feng G, et al. Salinity affects production and salt tolerance of dimorphic seeds of *Suaeda salsa*. Plant Physiology and Biochemistry 2015; 95:41-48. doi: 10.1016/j.plaphy.2015.07.005. PMID: 26184090.

[94] An Y-Q, Lin L. Transcriptional regulatory programs underlying barley germination and regulatory functions of gibberellin and abscisic acid. BMC Plant Biology. 2011;**2011**(11):105

[95] Finch-Savage WE, Leubner Metzger G. Seed dormancy and the control of germination. New Phytologist. 2006;**171**(3):501-523

[96] Han C, Yang PF. Studies on the molecular mechanisms of seed germination. Proteomics. 2015;**15**:1671-1679

[97] Betzen BM, Smart C, Maricle KL, MariCle BR. Effects of increasing salinity on photosynthesis and plant water potential in Kansas salt marsh species. Transactions of the Kansas Academy of Science. 2019;**122**(1-2):49-58

[98] Sheldon AR, Dalal RC, Kirchhof G, Kopittke PM, Menzies NW. The effect of salinity on plant-available water. Plant and Soil. 2017;**418**(1):477-491

[99] Mwando E, Han Y, Angessa TT, Zhou G, Hill CB, Zhang XQ. Genome-wide association study of salinity tolerance during germination in barley (*Hordeum vulgare* L.). Frontiers. Plant Science. 2020;**11**:118

[100] Flowers TJ, Gaur PM, Gowda CL, Krishnamurthy L, Samineni S, Siddique KH, et al. Salt sensitivity in chickpea. Plant, Cell and Environment. 2010;**33**(4):490-509

Chapter 8

Comparative Analysis of Molecular Allergy Features of Seed Proteins from Soybean (Glycine max) and Other Legumes Extensively Used for Food

Andrea Roman-Mateo, Esther Rodriguez-de Haro, Jose M. Berral-Hens, Sonia Morales-Santana and Jose C. Jimenez-Lopez

Abstract

Food allergies due to eating habits, pollution, and other factors are a growing problem in Western nations as well as developing countries. Symptoms of food allergies include changes in the respiratory and digestive systems. Legumes are a potential solution to the enormous demands for healthy, nutritive, and sustainable food. However, legumes also contain families of proteins that can cause food allergies. Some of these legumes include peanut, pea, chickpea, soy, and lupine. It has been shown that processing can alter the allergenicity of legumes since thermic and enzymatic resistance can affect these properties. Cross-reactivity (CR) is an allergy feature of some allergen proteins when the immune system recognizes part of the common share sequences (epitopes) in these allergic proteins. The research about molecular allergy includes comparisons of immunoglobulin E (IgE) and T-cell epitopes, assessment of three-dimensional structure and comparison of secondary structure elements, post-transduction modifications analysis by bioinformatic approach, and post-transduction modifications affecting epitopes properties may facilitate molecular tools to predict protein allergic behavior establishing prevention measurements that could promote the use of legumes and other seeds. This chapter provides an overview of the structural features of the main allergen proteins from legumes and their allergenic potential.

Keywords: food allergy, cross allergenicity, legumes, allergen proteins, soy, lupine

1. Introduction

Legumes are dicotyledon plants in the order Fabales and the family Fabacea. They produce fruit contained in pods and filled with seeds. In this chapter, we discuss three species of legumes in the genus *Lupinus* (*Lupinus albus*, *L. angustiflora*, and *Lupinus luteus*) and the most common allergenic species of the family Papilionaceae, including soja *Glycine max*; *Arachis hypogaea*, *A. duraensis*, and *A. ipaensis*; lentil (*Lens culinaris*); pea (*Pisum sativum*); and chickpea (*Cicer arietinum*).

A food allergy is an immune system reaction that occurs after eating certain types of food. Symptoms are variable and can be caused even by small amounts of allergenic proteins, leading to hives, swollen airways, and digestive problems. Food allergies are a growing concern worldwide. This increase is suspected to be related with industrial production, pollution, additives, and consumption of trash food [1]. There are reports of children of East Asian or African ethnicity in Western nations having an increased risk of developing food allergies compared with Caucasian children. This suggests that adopting Westernized food habits could increase food allergies in African or Asian countries [2, 3].

The research about healthy, low-cost alternative products that can meet the enormous demands of a growing population involve legumes [4]. Legume crops represent a sustainability solution, serving as a fundamental source of high-quality alternative protein, reducing the emission of greenhouse gases, allowing the sequestration of carbon in soils, saving the CO_2 print thanks to the nitrogen fertilizer, it free high-quality organic matter that facilitate water retention and perform the soil nutrients circulation among others uses [5]. Despite their advantages, legumes contain proteins that can potentially cause food allergies. Several allergens from different legumes have been identified and characterized as proteins with potential allergic effects. These include lentil, pea, chickpea, soy, peanut, and lupine [6].

Clinically, the absence of sensibilization phase is a reliable indicator of the tolerance to an allergen. In this context, the presence of sensitization to a specific allergen protein has to be proven [7] both, the specific reactivity to a particular allergen protein and the cross-reactivity to other related allergens. The most frequent cross-reactivity process described clinically is that between lupin and peanut [8].

In Spain, consumption of legumes is common because they are an important part of the Mediterranean diet. It is estimated that consumption of legumes in Spain is 4.8 kg per year, with a greater percentage of children eating them as compared to adults. Legume consumption in Spain is greater in girls than in boys [9]. One study in Spain showed that food allergies were detected in 20.8% of children and 14% of adults. In the overall Spanish population, legumes were responsible of the 14.3% of the food allergies [10]. Another study of Spain's pediatric population found that 10% of children suffered from food allergies caused by lentil and 6.7% of children suffered from food allergies caused by peanuts. Lentil was found to be the most allergenic, causing 78% of reactions, followed by chickpea (72%) and peanut (33%) [9].

In Europe, legumes are the fifth-leading cause of food allergies [11]. A meta-analysis of studies conducted in Europe between January 2000 and September 2012 found that the percentage of the population with symptoms of food allergies plus specific immunoglobulin E (IgE) positivity activation to at least one food allergen was 3%–4.6% in children and 2.2%–2.66% in adults [12]. The same study concluded that the frequency of food allergy is greatest in northwestern European countries compared to southern European countries, which had the lowest prevalence. Some factors related to food allergies include environmental, genetic, and epigenetic factors that could suggest differences between global populations [13].

The general prevalence of food allergies is not clearly defined due to the lack of reliable data and the highly variable allergy patterns in different parts of the world. A selection of mixed developed country data (Allergy, Asthma & Immunology Research

2018) found that some allergies, like those to peanut, demonstrate heritability in Caucasian populations; skin immune responses shows differences between Asians and Caucasians. These types of studies have not yet been conducted in non-White populations, however, there exists some interest data showing that Black South African children present a significantly lower prevalence of peanut allergy compared to children of mixed-race origin (Black and Caucasian) by unknown factors [13].

One interesting fact about cross-reactivity is that it could be caused by proteins that come from species that are taxonomically distant. Examples of these antigens are panallergens, which are proteins conserved by evolution due to their important defense, structural, and storage functions [7]. If a person has an allergy to cow milk proteins, they are also probably allergic to goat milk proteins [14]. In the case of legumes, cross-reactivity to more than one legume is often found in children [9].

Overall, allergic features of allergen proteins could be attenuated by thermic proteolytic denaturalization due to the modification of the quaternary protein structure where superficial epitopes of these proteins' antigenic regions can still develop some allergenicity reactions. Despite this, there are studies that also show resistance to thermic, chemical, and proteolytic denaturalization, with is a common characteristic in legumes [15]. Some examples of resistance to denaturalization include allergen proteins like Cupins, very stable storage proteins that include legumins (11 S) and vicilins (7 S), both containing two common β-barrel structures in their globular domain. These appear to be a relevant stable structural motif, confirming resistance to denaturation and proteolysis [16]. Lipid transfer proteins (LTPs) have resistance to pepsin and to chemical digestion [17]; PR-proteins have thermostable structure [10] allowing them staying unalterable at physiological temperature. This stability plays an important role in allowing allergen active protein fragments to pass to the gastrointestinal tract, causing a food allergy.

There is a large public database of allergenic legume proteins with several isoforms. The commonly shared partial epitopes and their conservation in the same family of proteins in different species could be helpful in designing possible strategies to prevent cross-reactivity.

The aim of this work is to carry out an exhaustive molecular and structural analysis of the most common allergenic legume proteins through bioinformatic approaches.

2. Materials and methods

2.1 Search of legume proteins sequences

We used the Allergome and UniProt databases to search for allergenic legume proteins for this study. The proteins chosen are characterized by having complete sequences and being in mature form. The search was carried out on the available species of lentil, pea, chickpea, soybean, and lupine (**Table 1A-E**).

2.2 Alignment of sequences

The complete and mature sequences of lentil (*Len c 3*, *Len c 3.0101*, and *Len c aglutinin*), chickpea (*Cic a 1, Cic a 3, Cic a 4, Cic a 6*), pea (*Pis s 2* (7 s vicilin), *Pis s 3* (LTP), *Pis s 3.0101*(LTP), *Pis s 6* (PR-protein, Pis S aglutin, Pis s albumin)), lupine (*Lup a 1, Lup a alpha conglutin, Lup a delta conglutin, Lup a gamma conglutin, Lup a 4, Lup an 1, Lup an 1.0101, Lup an 3, Lup an 3.0101, Lup an alpha conglutin, Lup an delta*

Species	Protein name	Protein type	UniProtKB
A.			
Soy allergen sequences			
Glycine max	Gly m 5	Profilin	C6T9L1 (C6T9L1_SOYBN)
	Gly m 5.0301	Profilin	P25974 (GLCB1_SOYBN
	Gly m 8	2 s albumin	C6SYA7 (C6SYA7_SOYBN)
	Gly m 8.0101	2 s albumin	P19594 (2SS_SOYBN)
B.			
Selected sequences of Lupinus			
Lupinus albus	Lup a 1	7 s vicilin	Q53HY0 (CONB1_LUPAL)
	Lup a alpha conglutin	11 s conglutin	Q53I54 (Q53I54_LUPAL)
	Lup a delta conglutin	2 s albumin	Q333K7 (Q333K7_LUPAL)
	Lup a gamma conglutin	Aspartic protease	Q9FEX1 (CONG2_LUPAL)
	Lup a 4	PR-protein	O24010 (O24010_LUPAL)
Lupinus angustifolius	Lup an 1	7 s vicilin	B0YJF8 (B0YJF8_LUPAN)
	Lup an 3	LTP	A0A1J7GK90 (A0A1J7GK90_LUPAN)
	Lup an 3.0101	LTP	A0A4P1RWD8 (A0A4P1RWD8_LUPAN)
	Lup an alpha conglutin	11 s globulin	F5B8V6 (CONA1_LUPAN)
	Lup an delta conglutin	2 s albumin	F5B8W8 (COND1_LUPAN)
	Lup an gamma conglutin	Aspartic protease	Q42369 (CONG1_LUPAN)
Lupinus luteus	Lup l 4	PR- protein	P52778 (L18A_LUPLU)
C.			
Selected sequences of Pea			
Pisum sativum	Pis s 2	7 s vicilin	P13915 (CVCA_PEA)
	Pis s 3	LTP	A0A158V755 (NLTP2_PEA)
	Pis s 6	PR-protein	P13239 (DRR1_PEA)
	Pis s agglutinin	Agglutinin	B5A8N6 (B5A8N6_PEA)
	Pis s albumin	Albumin	P08688 (ALB2_PEA)
D.			
Selected sequences of Chickpea			
Cicer arietinum	Cic a 1	7 s vicilin	Q304D4 (Q304D4_CICAR)
	Cic a 3	LTP	O23758 (NLTP_CICAR)
	Cic a 4	PR-protein	Q39450 (Q39450_CICAR)
	Cic a 6	11 s globulin	Q9SMJ4 (LEG_CICAR)
E.			
Selected sequences of Peanut			
Arachis hypogaea	Ara h 1	7 s vicilin	B3IXL2 (B3IXL2_ARAHY)
	Ara h 1.0101	7 s vicilin	P43238 (ALL12_ARAHY)
	Ara h 2.0101	2 s albumin	Q6PSU2–2 (CONG7_ARAHY)

Species	Protein name	Protein type	UniProtKB
	Ara h 2.0201	2 s albumin	Q6PSU2–3 (CONG7_ARAHY)
	Ara h 3	11 s globulin	A1DZF0 (A1DZF0_ARAHY)
	Ara h 3.0201	11 s globulin	Q9SQH7 (Q9SQH7_ARAHY)
	Ara h agglutinin	Agglutinin	P02872 (LECG_ARAHY)
	Ara h 5	Profilin	D3K177 (D3K177_ARAHY)
	Ara h 5.0101	Profilin	Q9SQI9 (PROF_ARAHY)
	Ara h 6	2 s albumin	A1DZE9 (A1DZE9_ARAHY)
	Ara h 6.0101	2S albumin	Q647G9 (CONG_ARAHY)
	Ara h 7.0101	2 s albumin	Q9SQH1 (Q9SQH1_ARAHY9
	Ara h 7.0201	2 s albumin	B4XID4 (B4XID4_ARAHY)
	Ara h 7.0301	2 s albumin	Q647G8 (Q647G8_ARAHY)
	Ara h 8	PR- 10 protein	B1PYZ4 (B1PYZ4_ARAHY)
	Ara h 8.0101	PR-10 protein	Q6VT83 (Q6VT83_ARAHY)
	Ara h 8.0201	PR- 10 protein	B0YIU5 (B0YIU5_ARAHY)
	Ara h 9.0101	9 k-LPT	B6CEX8 (B6CEX8_ARAHY)
	Ara h 10.0101	16kD protein	Q647G5 (OL101_ARAHY)
	Ara h 11.0101	14KD oleosin	Q45W87 (OL111_ARAHY)
	Ara h 11.0102	14kD oleosin	Q45W86 (OL112_ARAHY)
	Ara h 13.0102	Defensine	C0HJZ1 (DEF3_ARAHY)
	Ara h 14.0101	17.5kD oleosin	Q9AXI1 (OL141_ARAHY)
	Ara h 14.0102	17kD oleosin	Q9AXI0 (OL142_ARAHY)
	Ara h 14.0103	17kD oleosin	Q6J1J8 (OL143_ARAHY)
	Ara h 15.0101	17kD oleosin	Q647G3 (OLE15_ARAHY)
	Ara h 16	7 k LPT	A0A445DA28 (A0A445DA28_ARAHY)
	Ara h 17	11 k LTP	A0A445AL51 (A0A445AL51_ARAHY)
Arachis duranensis	Ara d 2	2 s albumin	A5Z1Q8 (A5Z1Q8_ARADU)
	Ara d 6	2 s albumin	A5Z1Q5 (A5Z1Q5_ARADU)
Arachis ipaensis	Ara i 2	2 s albumin	A5Z1Q9 (A5Z1Q9_ARAIP)
	Ara i 6	2 s albumin	A5Z1Q6 (A5Z1Q6_ARAIP)
F.			
Selected sequences of Lentil			
Lens culinaris	Len c 3	LTP	A0AT28 (NLTP1_LENCU)
	Len c 3.0101	LTP	A0AT29 (NLTP2_LENCU)
	Len c agglutinin	Agglutinin	P02870 (LEC_LENCU)

Table includes the species name, the common name of the allergen, the type of protein according to its biological nature/function, and the UniProt entry name (UniProtKB). All sequences were used for alignment, T-cell epitope search, and IgE analysis. Sequences from all lupin and soybean species were used for the post-translational modification search tasks (A and B). For secondary and tertiary structure assessment, only the sequences of interest were used: G. max *(Gly m 5, Gly m 5.0301, Gly m 8, and Gly m 8.0101);* L. albus *(Lup a 1 and Lup a alpha conglutin);* Lupinus angustifolius *(Lup an alpha e);* P. sativum *(Pis s albumin);* C. arietinum *(Cic a 6); and* A. hypogaea *(Ara h 5.0101).*

Table 1.
Summary of the sequences used in successive studies.

conglutin, Lup an gamma conglutin, Lup l 4), and peanut (*Ara d 2, Ara d 6, Ara h 1, Ara h 1.0101, Ara h 2, Ara h 2.0101, Ara h 2.0201, Ara h 2.0202, Ara h 3, Ara h 3.0201, Ara h agglutin, Ara h 5, Ara h 5.0101, Ara h 6, Ara h 6.0101, Ara h 7.0101, Ara h 7.0102, Ara h 7.0301, Ara h 8, Ara h 8.0101, Ara h 8.0201, Ara h 9.0101, Ara h 10.0101, Ara h 11.0101, Ara h 11.0102, Ara h 13.0102, Ara h 14.0101, Ara h 14.0102, Ara h 14.0103, Ara h 15.0101, Ara h 16, Ara h 17)* were aligned by pairs against soybean allergens (*Gly m 5, Gly m 5.0301, Gly m 8, Gly m 8.0101*) extracting the identity percentage and comparing the possible differences in the amino acid nature of the protein sequences (positive charge, negative charge, and polarity) of the allergens listed above.

2.3 Functional domain analysis

We used the program Pfam v34.0 (http://pfam.xfam.org/) to identify the possible domains present in the isoforms of legume proteins.

2.4 Post-translational modification site prediction

We used the MusiteDeep deep learning framework (https://github.com/duolinwang/MusiteDeep_web) to search for the presence of possible post-translational modifications and identify how they affect the potential allergenicity of the study proteins [18]. The prediction models used are phosphorylation (Y, S, T); N-linked glycosylation (N); O-linked glycosylation (S, T); ubiquitination; N6-acetyllysine (K); Methylarginine (R); Methyllysine (K); Hydroxyproline (P) and Hydroxylysine (K) with a threshold value of 0.8.

S-nitrosylations and T-nitrations were also studied via the iSNO-AAPair tool (Y. Xu et al., 2013), which was used to predict cysteine S-nitrosylation sites (http://app.aporc.org/iSNO-AAPair) with a threshold value greater than 0.8. The GPS-YNO2 tool (Liu et al., 2011) was used to predict tyrosine nitration sites (http://yno2.biocuckoo.org).

2.5 Secondary structure assessment

Secondary structure was assessed using PSIPRED (http://bioinf.cs.ucl.ac.uk/psipred/). Sequence alignment was performed with CLUSTALW (https://www.genome.jp/tools-bin/clustalw), which was visualized with the BioEdit program, and in which the consensus secondary structure was annotated.

2.6 Modeling of three-dimensional structure

The three-dimensional structures of olive ALDH proteins were modeled using the Phyre2 web program (http://www.sbg.bio.ic.ac.uk/phyre2), which is based on Markov algorithms to generate alignments of the problem protein sequences with proteins with experimentally obtained protein crystallographic models (PDB).

2.7 Identification of IgE-binding epitopes

We used the AlgPred server (www.imtech.res.in/raghava/algpred/submission.html), which creates arrays using sequences from known allergens, to identify IgE-binding epitopes and to determine potential allergenicity of proteins based on of their amino acid and dipeptide composition.

2.8 Identification of T cell binding epitopes

We used the ProPred program (Singh et al., 2011) (http://webs.iiitd.edu.in/raghava/propred/) to analyze the protein sequences of legumes in the study. The analysis was performed with a 2% threshold for the most common human HLA-DR alleles among the Caucasian population: [DRB1*0101 (DR1), DRB1*0301 (DR3), DRB1*0401 (DR4), DRB1*0701 (DR7), DRB1*0801 (DR8), DRB1*1101 (DR5), and DRB1*1501 (DR2)].

3. Results and discussion

3.1 Sequences obtained from the Allergome database

We used the Allergome database to retrieve the available sequences of complete proteins of legumes, following the link to UniProt. The legumes included in this study are lentil, lupin, pea, chickpea, and peanut. Only two major allergens (*Gly m 5* and *Gly m 8*) with their available isoforms were extracted from soybean and used as reference to carry out the alignments and further analyses.

The reference proteins, soybean major allergens *Gly m 5* and *Gly m 8* with their isoforms, correspond to profilin, 7 s globulins, and albumin 2 s protein families. The allergen *Gly m 8* is considered to have the highest sensitivity [19], specificity, and reproducibility [20] to clinical reaction to soybean in atopic patients. The combination of *Gly m 5* and *Gly m 8* was suggested as one of the best ways to perform the estimation of the sensitization level and to improve the diagnosis of soybean allergy in children [21]. Thus, in the case of high similarity between the sequences of these soy allergens and the allergens of the other legumes included in this study, the diagnosis of possible cross-reactions between them could be facilitated.

3.2 Alignment of allergen protein sequences

Sequence alignments were performed to compare the common and differential features between allergen proteins and legumes. Overall, and according to the CODEX Alimentarius Commission in 2003, only proteins with a percentage of identity greater than 50% by local alignment (BLAST) are at risk of allergy or cross-reactivity [22]. Therefore, results obtained from protein–protein alignment beforehand do not show values high enough to make a prediction of possible cross-reactivity between soybean proteins and the rest of the legumes (**Table 2**).

The highest percentage of identity was the result of the alignment between the *Gly m 5* proteins and the *Gly m 5.0301* isoform (**Table 3**) with the *Lup a 1* protein with values of 48.41% and 48.72%, respectively (**Table 2D**). However, these percentages do not exceed the minimum alignment percentage recommended as guidance. Despite this, there are reported cases of cross-reactivity between other proteins with which there is a percentage lower than the standard minimum value considered for cross-reactivity and lower than that which occurs between these proteins, as in the case of *Gly m 8* and *Ara h 2* [23], with an identity percentage of 31.46% (**Table 2F**).

The multiple alignment analysis between *Gly m 5* and the isoform *Gly m 5.0301* with the *Lup a 1* protein obtained a percentage of common identity of 35.80% with 207 identical positions (Image 1).

Glycine max	**Arachis duranensis**		***Lens culinaris***		
Protein name	**Ara d 2**	**Ara d 6**	**Len c 3**	**Len c 3.0101**	**Len c agglutinin**
A.					
Percentages of Amino Acid Sequence Identity by Alignment of Peanut and Lentil Species against Reference Soybean Sequences					
Gly m 5	8428	6067	5239	5157	11.803
Gly m 5.0301	9009	5909	4556	5817	11.349
Gly m 8	32.738	29.94	9942	10.465	9375
Gly m 8.0101	33.333	29.94	9942	9884	9278

Glycine max	***Lupinus angustifolius***						
Protein name	**Lup an 1**	**Lup an 10,101**	**Lup an 3**	**Lup an 30,101**	**Lup an alpha conglutin**	**Lup an delta conglutin**	**Lup an gamma conglutin**
B.							
Percentages of Amino Acid Sequence Identity by Alignment of Lupin Species against Reference Soybean Sequences							
Gly m 5	24.463	39.739	5843	4	17.304	8444	15.028
Gly m 5.0301	24.463	39.739	4.31	4	17.304	8444	14.657
Gly m 8	8114	6209	11.561	11.243	6616	35.62	5298
Gly m 8.0101	7877	6209	12.069	11.765	6616	36.25	5066

Glycine max	***Cicer arietinum***				**Arachis ipaensis**	
Protein name	**Cic a 1**	**Cic a 3**	**Cic a 4**	**Cic a 6**	**Ara i 2**	**Ara i 6**
C.						
Percentages of Amino Acid Sequence Identity by Alignment of Chickpea and Peanut Species against Reference Soybean Sequences						
Gly m 5	36.759	6378	8	13.587	8753	6292
Gly m 5.0301	37.575	6378	7556	8	8.85	6136
Gly m 8	7143	10.526	5021	7585	31.461	29.94
Gly m 8.0101	6513	10.526	5021	7585	31.461	30.539

Glycine max	***Lupinus albus***					
Protein name	**Lup a 1**	**Lup a 4**	**Lup a alpha conglutin**	**Lup a delta conglutin**	**Lup a gamma conglutin**	**Lup l 4**
D.						
Percentages of Amino Acid Sequence Identity by Alignment of Lupin Species against Reference Soybean Sequences						
Gly m 5	48.417	6.25	16.637	8036	13.645	6798
Gly m 5.0301	48.717	6.25	16.637	8259	14.098	7456
Gly m 8	5151	10.698	6501	35.625	4425	13.115
Gly m 8.0101	5009	10	6.18	36.25	4435	13.115

Glycine max	***Pisum sativum***					
Protein name	**Pis s 2**	**Pis s 3**	**Pis s 3.0101**	**Pis s 6**	**Pis S agglutin**	**Pis s albumin**
E.						
Percentages of Amino Acid Sequence Identity by Alignment of Pea Species against Reference Soybean Sequences						
Gly m 5	41.638	5467	5882	6798	9362	6798
Gly m 5.0301	41.638	5145	5369	6798	11.429	10.444
Gly m 8	5759	11.765	10.588	13.402	11.273	8.98
Gly m 8.0101	5.41	11.176	10.588	13.402	10.204	9388

Glycine max	***Arachis hypogaea***							
Protein name	**Ara h 1**	**Ara h 1.0101**	**Ara h 2**	**Ara h 2.0101**	**Ara h 2.0201**	**Ara h 2.0202**	**Ara h 3**	**Ara h 3.0201**
F.								
Percentages of Amino Acid Sequence Identity by Alignment of Peanut Species against Reference Soybean Sequences								
Gly m 5	36.585	35.726	8753	8811	8874	9031	15.412	14.685
Gly m 5.0301	36.748	35.885	8.85	9009	9292	9234	15.762	14.86
Gly m 8	5769	8307	31.461	32.738	34.818	33.133	5.41	6015
Gly m 8.0101	7329	7668	31.461	33.333	31.818	33.735	4.57	5636

Glycine max	***Arachis hypogaea***							
Protein name	**Ara h agglutinin**	**Ara h 5**	**Ara h 5.0101**	**Ara h 6**	**Ara h 6.0101**	**Ara h 7.0101**	**Ara h 7.0102**	**Ara h 7.0301**
G.								
Percentages of Amino Acid Sequence Identity by Alignment of Peanut Species against Reference Soybean Sequences								
Gly m 5	13.816	6349	6136	6606	6292	7982	7062	6292
Gly m 5.0301	13.717	4904	5.33	6951	6136	8296	6834	9131
Gly m 8	8571	10.734	6015	28.144	29.94	23.497	30.337	22,286
Gly m 8.0101	8571	10.674	9091	28.144	30.539	23.497	30.899	22.857

Glycine max	***Arachis hypogaea***							
Protein name	**Ara h 8**	**Ara h 8.0101**	**Ara h 8.0201**	**Ara h 9.0101**	**Ara h 10.0101**	**Ara h 11.0101**	**Ara h 11.0102**	**Ara h 13.0102**
H.								
Percentages of Amino Acid Sequence Identity by Alignment of Peanut Species aAgainst Reference Soybean Sequences								
Gly m 5	6181	7761	6.92	3596	7761	6982	7207	3139
Gly m 5.0301	6935	7539	6.92	3.82	6828	6982	7207	3139

Glycine max	*Arachis hypogaea*							
Protein name	**Ara h 8**	**Ara h 8.0101**	**Ara h 8.0201**	**Ara h 9.0101**	**Ara h 10.0101**	**Ara h 11.0101**	**Ara h 11.0102**	**Ara h 13.0102**
Gly m 8	11.429	10.233	11.64	10.405	6478	6.14	6.14	9877
Gly m 8.0101	11.792	11.64	11.64	10.405	6883	6.14	6.14	9259

Glycine max	*Arachis hypogaea*					
Protein name	**Ara h 14.0101**	**Ara h 14.0102**	**Ara h 14.0103**	**Ara h 15.0101**	**Ara h 16**	**Ara h 17**
I.						
Percentages of Amino Acid Sequence Identity by Alignment of Peanut Species against Reference Soybean Sequences						
Gly m 5	8744	7848	8296	7221	4698	3905
Gly m 5.0301	8744	7848	8296	7221	4698	4121
Gly m 8	5785	5859	5785	5.6	11.111	11.243
Gly m 8.0101	5372	5859	5785	5.6	11.31	11.243

Degree of identity resulting from the alignment of amino acid sequences. These have been obtained by alignment between soybean proteins, used as reference, against different legume species (lentil, chickpea, pea, lupine, and peanut) including major allergens and isoforms.

Table 2.
Percentages of amino acid sequence identity by alignment of different legume species against reference soybean sequences.

Alignment Frequency Calculations			
Average of the difference of the frequencies between the different isoforms of soybean proteins with the alignment of the different proteins of legume species.			
Gly m 5/Gly m 5.0301	0,599 (over all)	**values > 3%**	5587 (Cic a 6) 3646 (Pis s albumin)
Gly m 8/Gly m 8.0101	0,468 (over all)	**values > 3%**	3076 (Ara h 5.0101)
Max identity values obtained by sequences alignment			
Greater value	48,717 (over all)		
	Gly m 5.0301 vs. Lup a 1		

Table 3.
Summary of the largest (greater than 3%) and smallest differences as a result of legume–soy protein alignment.

These data show that the percentage of identity of allergens must be kept in mind to compare allergens and to predict potential allergenicity and cross-reactivity, since not only do sequential epitopes have to be taken into account for that purpose, but also 3D and specific structural conformations of particular allergen proteins must be considered.

Using the information obtained by alignment, some of the proteins in the comparative analysis with soybean could be of interest at the molecular allergy level, such as Lup a delta conglutin and Lup an delta conglutin with percentages of identity with *Gly m 8* and *Gly m 8.0101* ranging from 35 to 36%. It also presents notable alignment percentage differences with *Gly m 5* and *Gly m 5.0301* (**Table 2B, D**), with approximately 8% being the most notable difference in identity with respect to the other conglutins. Another candidate protein for analysis is *Lup a delta conglutin* with percentages of identity of 35.63% and 36.25% compared to *Gly 8* and its isoform *Gly m 8.0101*, respectively (**Table 2D**) and *Lup an delta conglutin* of 35.62% and 36.25%, respectively (**Table 2B**). The identity ratios are lower than the minimum value considered to establish cross-reactivity with soybean. However, with such similar percentages among conglutin sequences it is worthy to conduct a deeper analysis. Multiple alignment shows a high rate of conservation between lupin proteins from the different species of *L. albus* and *Lupinus angustifolia*. Comparison of gamma conglutin protein sequences of both species, soybean obtained a low identity percentage of 13–15% compared to *Gly m 5* and 4–5% compared to *Gly m 8* (**Table 2B, D**). Alignment between both conglutins showed an identity of 84.21%, with 128 identical positions and 12 similar positions (**Figure 1**), with an identity value high enough to consider cross-reactivity among them. Indeed, these sequences showed high conservation rate among lupin proteins from different species such as *L. albus* and *L. angustifolia*. The three-dimensional structure of these conglutins will be further analyzed in later sections (**Figure 2**).

Considering the identity percentages previously indicated, the Ara h 2 identity percentage of 31% at *Gly m 8* with demonstrated cross-reactivity and the 48% identity of *Lup a 1* with soybean, we found more cases of proteins with intermediate values. Such is the case of *Pis s 2* with *Gly m 5* and its isoform with an identity of 41.638% (**Table 2E**) and *Cic a 1* with 36.76% and 37.58% identity with *Gly m 5* and its isoform, respectively (**Table 2C**). On the other hand, the characterization of demonstrated cross-reactivity between soybean and peanut, as is the case of *Ara h 1* with *Gly m 5* and its isoform *Gly m 5.0301,* showed a 36.59% and 36.75% identity, respectively [24]. The rest of the alignments show percentages less than the described data of identity range and may be discarded from the depth in their CR study (**Table 2**).

Interestingly, the percentage of alignment identity between soybean isoforms was low, with values less than 1%, specifically, in the alignment of soybean major allergen *Gly m 5* and its isoform *Gly m 5.0301.* The sequences of these two allergens were compared to the rest of the legume proteins considered in this study. We obtained a different percentage of identity of 0.6%, as well as 0.47% when compared *Gly m 8* with *Gly m 8.0101* (**Table 3**). The largest differences were found between soybean isoforms and legumes; *Gly m 5/Gly m 5.0301* was 5.60% against chickpea protein *Cic a 6* (**Table 2C**); 3.65% against pea *Pis s albumin* (**Table 2E**) protein; and *Gly m 8 /Gly m 8.0101* 3.07% against peanut (*A. hypogaea*) protein *Ara h 5.0101* (**Table 2G**). **Table 3** summarizes this data.

The existence of differences between isoforms of other legume species of the same allergen protein family could open the way for new studies finding significant differences in multiple cross-reactivity candidacy. For example, such as the case of *Lup an 1* and *Lup an 1. 0101* with identity differences exceeding 13% in alignment with *Gly m 5*, and ranging between 24.46% and 39.74%, respectively (**Table 2B**). These differences make *Lup an 1* an unsuitable candidate for cross-reactivity, whereas its isoform *Lup an 1.0101* could be a candidate for cross-reactivity with soybean.

3.3 Post-translational modification analysis

Post-translational modifications affecting the allergen protein sequences have been defined and involved in processes like alcohol or tiol addition (glycosidations), methyl

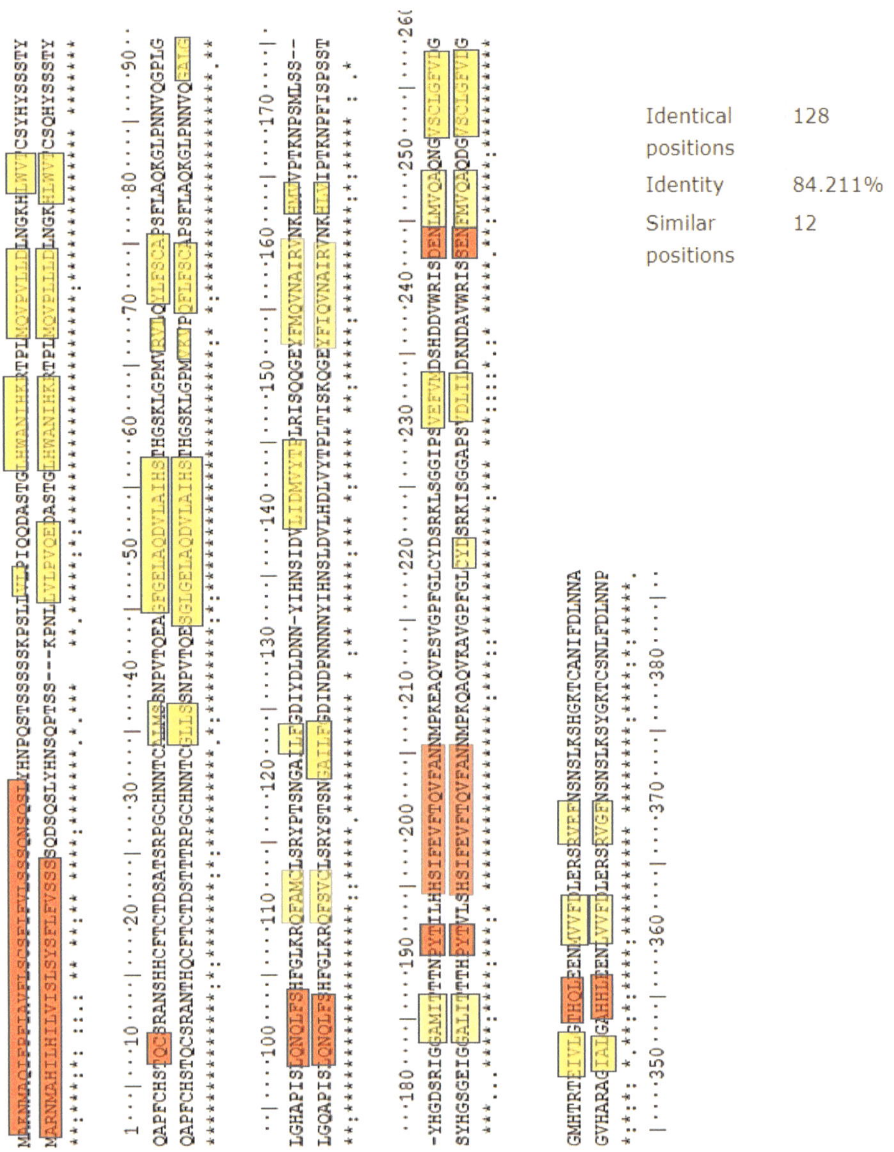

Figure 1.
*2D structure of allergen proteins. Multiple alignment of the major Lup a gamma conglutin (*Lupinus albus*) against Lup an gamma conglutin (*Lupinus angustifoluis*) with the secondary sequence represented in yellow by coil zones and in red by helix zones. In addition to the percentage of joint identity, number of identical amino acid positions and number of amino acid have similar physicochemical nature.*

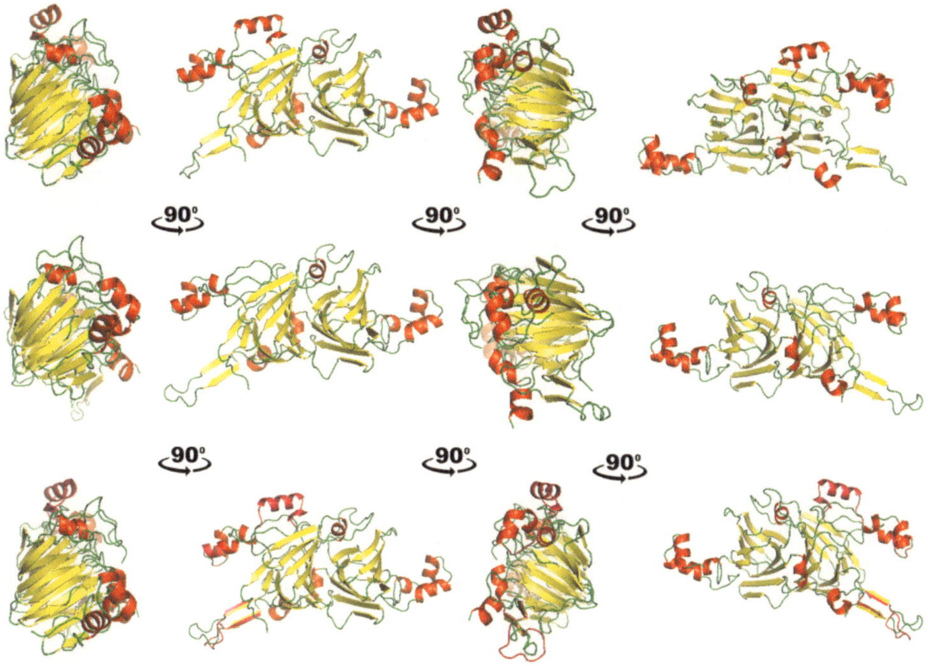

Figure 2.
Three-dimensional structural analysis of seed allergen proteins. Figures of first row corresponding to the 3D structures of the Lup a gamma conglutin protein; second row represent different views of Lup an gamma conglutin; and third raw are the figures of the consensus sequence with depicted match regions in pink color over the consensus figure (last row). Red color highlights the alpha-helix and yellow color the beta-strand.

groups (methylations), phosphates (phosphorylations), carboxyl groups (carboxylations), nitro groups (T-nitrations), or nitroxil groups (S- nitrosylations).

These types of modifications may induce rearrangements in structure, which could indirectly affect lineal and/or conformational epitopes' influence pm molecular allergy, limiting or favoring immunological recognition as well as generating antigenic diversity [25]. It is interesting to analyze location of where these modifications may occur and the type of modification together with the influence of these modifications in the 2D structural elements.

Phosphorylation is considered a factor of change of molecular pH dynamics [26], generating important alterations in the biophysics of the protein [27]. It has been observed sites of phosphorylation in most of the proteins examined: *Gly 5, Gly 8* and their isoforms; *Lup a 1, Lup a* alpha and delta conglutins (*L. albus*); *Lup an 1* and its isoform *Lup an 1.0101, Lup an alpha, Lup an delta* and *Lup an gamma (L. angustifolius)*. In the sequences of *Lup l 4 (L. luteus)* and *Cic a 6 (C. arietinum)* are also abundant modifications as glycosidations which potential importance in the allergenicity behavior of these proteins. In this regard, it has been demonstrated in some cases the increasing immunogenicity [28] for *Gly 5* and *Gly 8*; *Lup a 1, Lup a 4, Lup a alpha, delta*, and *gamma conglutins*; *Lup an 1* and it isoform *Lup an 1.0101, Lup an alpha* and *gamma conglutins*; *Lup l 4* and *Cic a 6* (**Table 4**).

Methylations are quite less abundant modifications. It is observed that their deficiency generates serious alterations in the functioning of proteins, thus having important implications on their three-dimensional structuring as carboxylation [29]. Only two methylation sites were found: one on *Lup a alpha conglutin* and one on *Lup an alpha conglutin* (**Table 4B**). Carboxylations were found on the *Gly m 8.0101* isoform;

Lup a alpha, delta, and *gamma conglutins*; *Lup an 1* and its isoform *Lup an 1.0101;* and *Lup an 3* and *Lup an alpha conglutin* (**Table 4A, B**).

Nitrosylation and nitrations generate strong covalent bonds in the protein structure [30, 31]. Nitrations were found on *Lup a 1, Lup a 4*, and *Lup a alpha conglutin*; *Lup a gamma conglutin*, *Lup an 1,* and *Lup an 1.0101*; *Lup an 3.0101*, *Lup an alpha* and *gamma conglutin*; *Lup l 4*; *Cic a 6* and *Ara h 5.0101*. Nitrosylations in comparison were less abundant, found in *Lup a alpha conglutin*; *Lup an 3* and its isoform *Lup an 3.0101*, and *Lup an alpha, delta,* and *gamma conglutins* (**Table 4**).

Post-translational modifications on T-cell epitopes have been found in *Gly m* 5.0301 isoform, a glycosidation at position 351, and a nitration at 172; *Lup a alpha conglutin* presents three methylation sites at positions 199, 448, and 497; *Lup a delta conglutin* contains a glycosidation site at position 76; a nitrosylation site at position 13 was found in *Lup an 3*, while in its isoform a nitration at position 104 and a nitrosylation at position 112 are highlighted; *Lup an delta* conglutin presents a candidate phosphorylation site at position 76 and Cic a 6 a nitrosylation at 107. In other cases, IgE epitopes are affected, with the only case of *Lup a alpha* conglutin with a methylation site at position 102. **Table 5** presents a summary of this data.

The direct implications of these post-translational modifications may be directly linked to the effects on the variation of the structure of these regions, generating differential epitopes recognition and consequently the allergen response.

Analyzing the location and type of modifications could help to elucidate the relationship of protein structure epitope distribution to the allergen potential of the protein, however, it will not be confirmed whether the different modifications would accentuate or lessen the allergenic impact until a clinical review of the process is carried out. The possibility of inducing post-translational modifications on plant proteins as a therapeutic tool is being examined [27].

3.4 Secondary structure analysis

The combined analysis of secondary structure with multiple alignments allows a direct sequence–structure–functional comparation between different allergen proteins. An interesting analysis has been made to identify the areas of allergens with shared mutual domains as part of structural domains with important implications for cross-reactivity potential.

The *Gly m 5, Gly m 5.0301*, and *Lup a 1* secondary structure comparison showed that in sequences of these proteins (**Table 2A**), the percentage of identity with *Lup a 1* was the highest compared to the rest of the alignments performed (**Table 3**). However, the percentage was not potentially enough to induce cross-reactivity. Comparative analysis between the secondary structure predictions of these proteins shows strong similarities in the distribution of α-helix and β-strand over middle regions of the proteins (amino acids 20–430) (**Figure 3**), giving an additional perspective of the possible regions with potential cross-reactivity in addition to the information provided by the alignments.

The three allergen proteins include Cupin superfamily domains with a wide variety of representative enzymes, but notably contains the non-enzymatic seed storage proteins [32]. Functional domains that could be candidates to potentially undergo post-translational modifications for *Lup a 1* are one of the two barrel domains with anti-parallel b-sheets. The first one is a Cupin_1.1 (**Table 6A**), a candidate for glycosidation (**Table 4B**). Similarly, in the case of *Gly m 5* and its isoform *Gly m 5.0301*, in both proteins where also present these modifications in their globular

Allergen	Post-translational modifications					
	Phosphorylation	**Glycosylation**	**Pyrrolidone carboxylic acid**	**Methylation**	**Nitration**	**Nitrosylation**
A.						
Post-translational modifications predicted over soybean: Glycine max (Gly m)						
Gly m 5	232; 234; 235	351	—	—	158;172	—
Gly m 5.0301	232; 234; 235	351	—	—	158; 172	—
Gly m 8	155; 156	120	—	—	—	14
Gly m 8.0101	155; 156	120	25	—	—	14
B.						
Post-translational Modifications Predicted Over Lupinus: Lupinus albus (Lup a), L. angustifolius (Lup an), and L. luteus (Lup l)						
Lup a 1	71; 79; 104	444	—	—	269;316	—
Lup a 4	—	13; 82	—	—	157; 269; 316	—
Lup a alpha conglutin	347	403	29	102	199; 448; 497	36; 334
Lup a delta conglutin	75;76	73; 108	27	—	—	—
Lup a gama conglutin	—	133	28	—	261	—
Lup an 1	80;82;85	152; 434	126; 158	—	340	—
Lup an 1.0101	80;82;85; 469; 488	434; 519	126; 158	—	340; 488	—
Lup an 3	—	—	23	—	—	13; 27
Lup an 3.0101	—	—	—	—	104	28; 112
Lup an alpha conglutin	247; 259; 341	397; 439	24	97	84; 442; 491	31
Lup an delta conglutin	76; 77; 80;83	—	—	—	—	42
Lup an gamma conglutin	357	130	—	—	259	350; 391; 440
Lup l 4	112	78; 82	—	—	100; 156	—
C.						
Post-translational Modifications Predicted Over Chickpea: Cicer arietinum (Cic a) and Peanut: Arachis hypogaea (Ara h)						
Cic a 6	139; 195; 207; 225; 271	1; 220	—	—	443	64; 107
Ara h 5.0101	—	—	—	—	6; 125	115

Specific amino acids affected by each type of post-translational modification on the different legume proteins: phosphorylation, glycosylation, carboxylation (pyrrolidone carboxylic acid), methylation, nitrosylation, and nitration sites. The (–) symbol means no results.

Table 4.
Post-translational modifications predicted over legumes.

Allergen name	Post-translational Modifications				
	Phosphorylation	**Glycosylation**	**Methylation**	**Nitration**	**Nitrosylation**
A.					
T-cell epitopes from allergens affected by post-translational modifications					
Gly m 5.0301		FVVNATSNL (351)		YLQGFDHNI (172)	
Lup a alpha conglutin			FGPLRRCN (199)		
			YVLNGSAWF (448)		
			YVAFKTNDI (497)		
Lup a delta conglutin		LVAALVLVV (76)			
Lup an 3					VLICMVVVS (13)
Lup an 3.0101				YKISTSTNC (104)	YKISTSTNC (112)
Lup an delta conglutin	LVVHTSASR (76)				
Cic a 6					FGMVFPGCV (107)
B.					
IgE epitopes from allergens affected by post-translational modifications					
Lup a alpha conglutin			IETWNPNNQEFECAG (102)		

This table summarizes the T-cell and IgE epitopes directly affected by the main post-translational modifications indicating the amino acid number affected.

Table 5.
T-cell and IgE epitopes from allergens affected by post-translational modifications.

domain (antiparallel β-barrels) (**Table 6A**), which is a candidate to undergo glycosylation (**Table 4A**). In three cases, modifications by glycosidation of one of their functional domains is a shared functional and allergenic feature.

Lup a gamma conglutin and *Lup an gamma conglutin* were analyzed. Although they belong to different species of lupin, they showed few differences in alignment and their comparison with soybean proteins of reference (**Table 2B, D**). The identity percentage among them is greater than 50%. These allergen proteins could be considered to exhibit CR, due to sequence identity but also to similarities of their secondary structure (**Figure 1**).

Regarding the predictions of post-translational modifications of these proteins relevant to 2D structural domains, it was found that *Lup a gamma conglutin* can be modified by a potential glycosidation (**Table 4B**). This modification is located in the

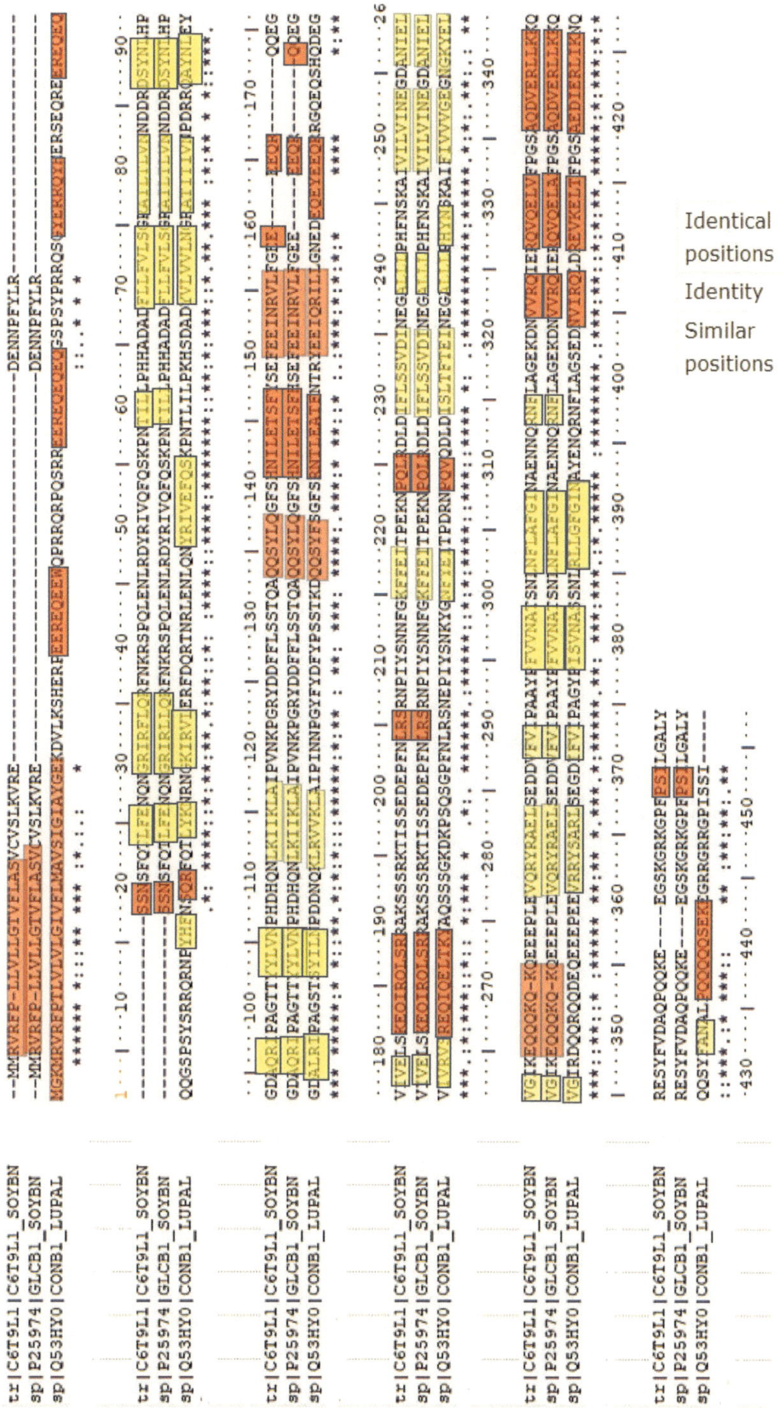

Identical positions	207
Identity	35.813%
Similar positions	157

Figure 3.
2D structure of allergen proteins. Multiple alignment of the major allergen Gly m 5, its isoform Gly m 5.0301 from (Glycine max) *and Lup a 1* (Lupinus albus) *together with the secondary sequence is represented in yellow by coiled-coil zones and in red by helix zones. In addition to the percentage of joint identity, number of identical amino acid positions and number of amino acid have similar physicochemical nature.*

Protein	Functional domain	Alignment amino acid range
A.		
Functional Domains Predicted Over Gly m 5, Gly m 5.0301 and Lup a 1		
Lup a 1	Cupin_1.1	332–486
	Cupin_1	137–227
Gly m 5	Cupin_1	240–389
	Cupin_2	86–144
Gly m 5.0301	Cupin_1	240–393
	Cupin_2	86–144
B.		
Functional Domains Predicted Over Lup a gamma conglutin and Lup an gamma conglutin		
Lup a gamma conglutin	Xylanase inhibitor C-terminal	271–428
	Xylanase inhibitor N-terminal	66–240
Lup an gamma conglutin	Xylanase inhibitor C-terminal	269–429
	Xylanase inhibitor N-terminal	63–237

This table summarizes the protein domains of the different proteins in their different types, specifying the range of amino acids that occupy in alignment.

Table 6.
Functional domains predicted over legumes allergens.

region of the protein domain xylanase inhibitor C-terminal (**Table 6B**). *Lup an gamma conglutin* has two possible domains affected by post-translational modifications: a phosphorylation and two nitrosylations (**Table 4B**) that affect the region comprised in the C-terminal xylanase inhibitor domain (**Table 6B**) and two nitrosylations (**Table 4B**) over the same domain. It also presents a glycosidation (**Table 4B**) in the xylanase inhibitor N-terminal domain (**Table 6B**).

3.5 Three-dimensional structure analysis

Analysis of three-dimensional structure of proteins (**Figure 4**) provides insight into their sequence conformation and epitope arrangement. It also helps to determine the consequences of possible structural changes occurring between protein isoforms with minimal or large number of changes (**Table 2**) in their sequences [33].

Post-translational modifications over protein domains also may generate changes in their three-dimensional structure, affecting exposure epitopes and increasing or decreasing their allergenic potential.

Some candidates to examine the three-dimensional structure are *Gly m 5*, *Gly m 5.0101*, and *Lup a 1* that share common barrel domains with alternating folds between the α-helix and β-strand. These domains are in a special conformation, forming a solenoid in which the β-strand is arranged on the inside of the toroid and the α-helix is arranged on the outside in the same domain (**Figures 2** and **5**).

The structural differences observed in the consensus structure between the three structures indicate that in *Gly m 5.0301*, an element of the 2D structure corresponding to a β-strand structural connection is not present in the isoform *Gly m 5*. Neither is it present in *Lup a 1*, which is a specific and important structural feature that can make a

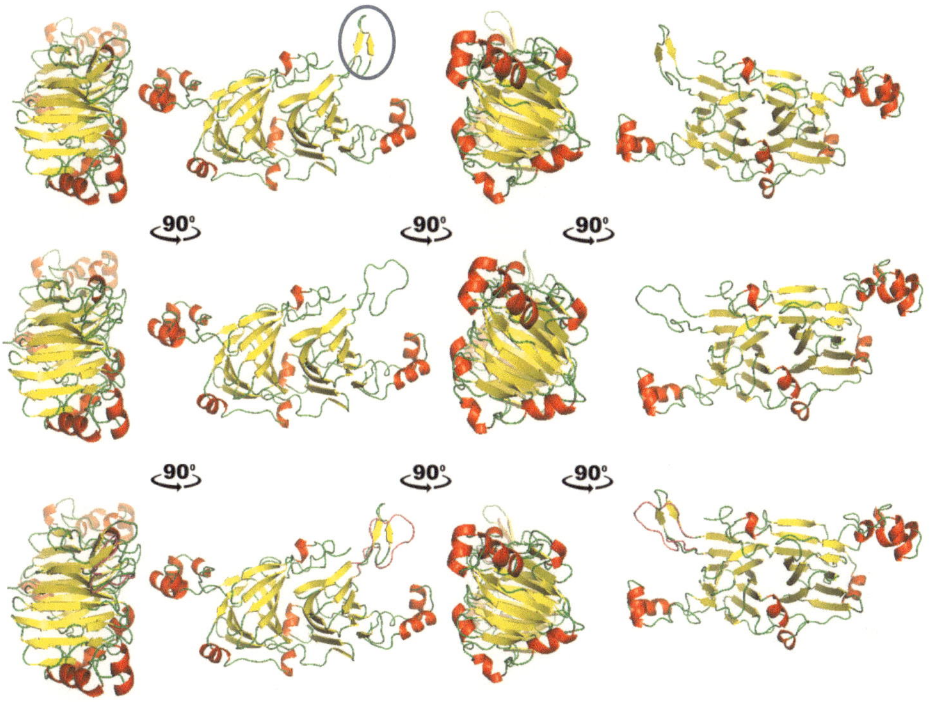

Figure 4.
3D structural analysis of seed allergen proteins. Three-dimensional structures of the Gly m 5.0301 *proteins are described, followed by* Gly m 5.0301 *and the change points between the two proteins marked in soft pink color in consensus figure (last row). Red denotes the alpha-helix and yellow denotes the beta-strand. T-epitope location is marked by a blue circle.*

specific conformational epitope (**Figures 4** and **5**). This structural change does not contain any epitope sequence. However, the change found is located between the *Cupin-1* domain of *Gly m 5* and its isoform, whereas this change in *Lup a 1* is located in the Cupin_1.1 domain (**Table 6A**).

Tridimensional structure comparison between Lup a gamma conglutin and Lup an gamma conglutin result on two principal differences observed between both conglutins, which is an α-helix in the gamma conglutin of *L. albus* that is not present in *L. angustifolius* (**Figure 2**). Regarding post-translational modification sites, in this loop there are no predicted modifications in this region encompassing the N-terminal xylanase inhibitor domain (**Table 6B**).

The 3D analysis was useful to determine other cases of interest previously mentioned, such as *Pis s 2* and *Cic a 1* in comparison with *Gly m 5* and its isoform that showed considerable identity ratios (**Table 2C, E**). *Lup an 1* and *Lup an 1.0101* showed large differences between their identity, and even more differences were found when compared to *Gly m 5*, which is somehow reflected in their 3D structures.

3.6 Identification and analysis of T-cell binding epitopes

An epitope is the portion of a macromolecule that is recognized by the immune system, specifically the sequence to which antibodies, B-cell receptors or T-cell receptors, can bind to initiate an immune response. Analysis of the epitopes shared for specific allergen proteins could be relevant to identify potential cross-reactivity.

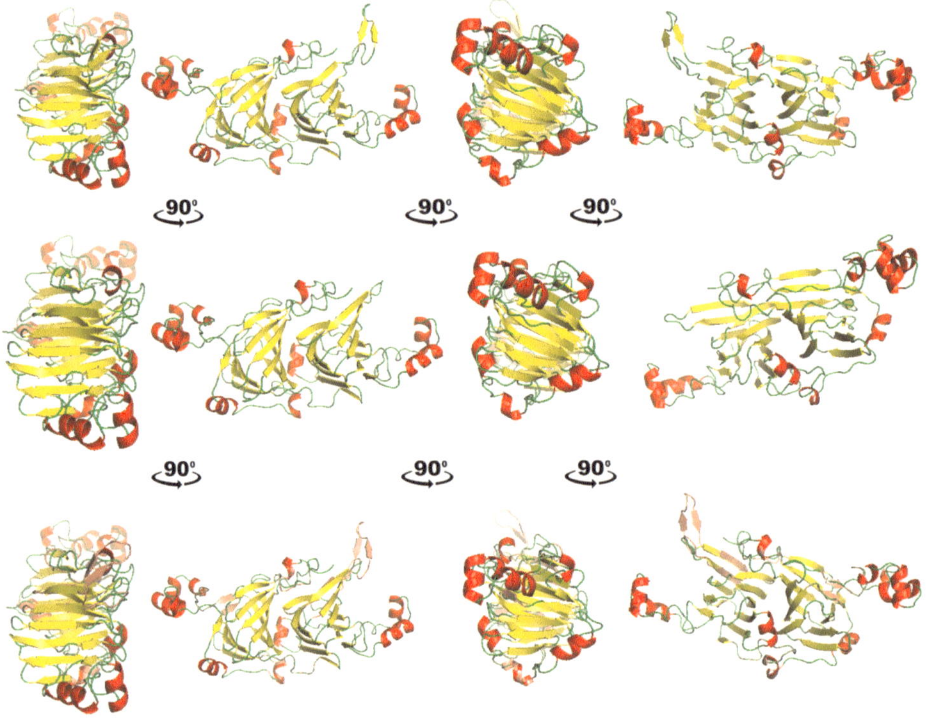

Figure 5.
3D structural analysis of seed allergen proteins. Three-dimensional structures of the Gly m 5 *proteins followed by* Lup a 1 *and representative changes between these two proteins marked in pink in the consensus figure (last row). Red denotes the alpha-helix and yellow denotes the beta-strand.The three-dimensional structure of the proteins* Gly m 5, Gly m 5.0301 *(*Glycine max*), and* Lup a 1 *(*L. albus*) showed a structure with large number of similarities, which is also reflected in the previous analysis of their secondary structure (**Figure 3**), with two barrel domains common in all of them.*

Presence of common T-cell epitopes among different legume species may support cross-reactivity processes; the greater the probability of occurrence, the larger the number of common epitopes.

The data obtained from the analysis of T-cell epitopes allows us to know which epitopes are shared among allergen proteins in the different legume species and to examine possible cases of cross-reactivity. Thus, in the case of soybean *G. max*, epitopes common to peanut, *A. hypogaea* species and chickpea *C. airietinum* species are described in **Table 7A**. It is remarkable that the soybean protein isoform *Gly m 5.0301* has an epitope in common with *Ara h 9.0101*, while the major allergen *Gly m 5* does not contain this epitope (**Table 7A**). This feature may be related to the cross-reactivity between specific sequences and these legume cultivars containing these specific proteins.

On the other hand, the different lupin species show that up to 18 T-cell epitopes are found commonly shared between *L. albus* and *L. angustifolius* (**Table 7B** part 1, 2, 3 and 4). Shared epitopes are also observed between *L. albus* and *A. hypogaea* (four epitopes) (**Table 7B** part 1, 2 and 4); *A. duranensis* (one epitope), *C. arietinum* (same number of epitopes) (**Table** 7 part 1). Comparison with *L. angustifolius* showed three epitopes commonly shared with *A. hypogaea* (**Table 7B** parts 2, 3 and 4), and one epitope with *C. arietinum* and *L. culinaris* (**Table 7B** part 3).

Among these allergen proteins, there are also epitopes shared more than one time among more than two species. The same epitope is shared among the allergenic proteins: *Lup a 4* with *Ara h 8.0101* and *Cic a 4* (**Table 7B** part 1); *Lup an alpha conglutin*, *Lup an 3.0101, Ara h 3,* and *Ara h 3.0201* (**Table 7B** part 4). the most shared epitope was between *Lup an 3*, *Lup an 3.0101*, *Ara h 9.0101*, *Ara h 17*, *Cic a 3*, *Len c 3*, and *Len c 3.0101* (**Table 7B** part 3).

Prediction of secondary and tertiary structures allowed us to determine the spatial location of epitopes in proteins and to assess whether they may be affected in their spatial arrangement by post-translational modifications in protein domains over interest proteins.

Gly m 5, *Gly m 5.0301*, and *Lup a 1* analysis also showed that T-epitope regions founded over these proteins integrate part of the functional barrel domains of these proteins. In the case of *Gly m 5*, a single T-epitope (**Table 6A**) is located in the region of the structural domain between β-strands (**Figure 5**). This region is located into Gly m 5-barrel domain (Cupin_1) (**Table 6A**) in the amino acidic region located close to the site of glycosidation (**Table 5A**). This structural epitope is of special interest by its specificity, location, and potential specific allergenicity induced by this protein.

The T-cell epitopes analyzed on *L.* gamma conglutins resulted in the presence of two epitopes on the C-terminal xylanase and one on the N-terminal xylanase domain of *L. albus* (**Table 6B**, **Table 7B** part 1and 2) and one over N-terminal xylanase domain of *L. angustifolius* (**Table 6B** and 7 part 1). These are not directly or proximally affected by post-translational modifications, but they do affect the domains in which they are located.

Therefore, epitopic regions matched between *L. albus* and *L. angustifolius* conglutin, which are the most abundant compared to other epitopes (**Table 7B**). This supports the idea of conservation of protein structures and evidences the data found by simple comparative alignment.

3.7 Identification and analysis of IgE-binding epitopes

The IgE antibodies are produced by immune B cells, which in turn are stimulated by T cells responsible for recognizing the epitope in a sensitization step. To trigger the allergen inflammatory process, IgE antibodies stimulate the release of histamines. Thus, the recognition of these sequences allows for predicting the recognition capacity of IgE antibodies and whether they will potentially trigger the allergenic response (**Figure 6**).

The analysis of the allergenic nature of the protein based on amino acid and dipeptide analysis composition has been used for the assessment of the above proteins. It is noticeable that the 30cases with clinically confirmed allergenic epitopes are predicted by their sequence to have an allergenic nature, as is the case of *Gly m 8* (**Table 8B**), *Ara h 13.0102*, and *Ara h 15.0101* (**Table 8**: D). Other potential allergens are *Lup a 4* (**Table 8A**), *Lup an 3* and *Lup an 3.0101* (**Table 8A**) and *Lup an delta conglutin*; *Pis s 3, Pis s 3.0101, Pis s 6, Pis s agglutin* and *Pis s albumin* (**Table 8B**); *Ara h 5, Ara h 5.0101* (**Table 8C**), *Ara h 8, Ara h 8.0101, Ara h 8.0102* (**Table 8D**); as both: 43 *Lup l 4* (**Table 8A**); *Ara h 17* (**Table 8D**) and *Cic a 3* (**Table 8C**).

Other proteins assessed as ambiguous or non-allergenic even though they present bibliographic and clinical antecedents of being allergenic include *Lup a gamma* conglutin [34] and *Lup an gamma conglutin* [35] (**Table 8A**); *Ara h 10.0101* [36], *Ara h 11.0101*, and *Ara h 11.0102* [37]; and *Ara i 2.0101* and *Ara i 6.0101* [38] (**Table 8C**).

Allergen name	T-cell epitopes	
	LRSSNSFQT	LRSRNPIYS
A.		
Range of amino acids occupied by T-cell epitopes joint over soy		
Gly m 5		288–296
Gly m 5.0301	36–44	242–250
Ara h 9.0101	21–29	
Cic a 1		250–258

Allergen name	T-cell epitopes					
	LVLVLGIVF	MMACNGLTI	YVLHKIEEI	FVLSSSQNS	LVAALVLVV	LVVHTSASR
B part 1						
Range of amino acids occupied by T-cell epitopes joint over lupin, peanut, and chickpea						
Lup a 1	11–19					
Lup a 4			66–75			
Lup a alpha conglutin						
Lup a delta conglutin					67–75	73–81
Lup a gamma conglutin		16–24		63–71		
Lup an delta conglutin					62–70	69–77
Lup an gamma conglutin		13–21		77% (FVSSSSQD) 69–77		
Ara d 6	13–20					
Ara h 8.0102			77% (YVLHKIDAI) 66–74			
Cic a 4			88% (YVLHKIEAI) 123–132			

Allergen name	T-cell epitopes					
	FQRLNALEP	LRCAGVALS	IRVLERFDQ	FGPLRRCN	VVLNGRATITI	IVRNIKGKN
B part 2						
Lup a 1			133–138		177–190	
Lup a 4						
Lup a alpha conglutin	83–91	112–120		192–200		279–287
Lup an 1			80% (IRVLERFNQ) 204–212		248–259	

Allergen name	T-cell epitopes					
	FQRLNALEP	**LRCAGVALS**	**IRVLERFDQ**	**FGPLRRCN**	**VVLNGRATITI**	**IVRNIKGKN**
Lup an 1.0101			80% (IRVLERFNQ) 204–213		248–260	
Lup an alpha conglutin	86–94	115–123				286–294
Lup an delta conglutin				191–198		
Ara h 1			80% (IRVLQRFDQ) 204–212			
Ara h 1.0101			80% (IRVLQRFDQ) 193–201			

Allergen name	T-cell epitopes			
	IVRVSREQI	**IRVNKHM**	**VRRVRRPH**	**WRISDEN**
B part 3				
Lup a 1	302–310			
Lup a alpha conglutin			355–363	
Lup a gamma conglutin		318–326		412–420
Lup an 1	77% (IVRVSKKQI)373–381			
Lup an 1.0101	77% (IVRVSKKQI) 373–381			
Lup an 3.0101			360–367	
Lup an delta conglutin		88% (IRVNKHL) 324–332		88% (WRISSEN) 421–429

Allergen name	T-cell epitopes					
	FPILGWLGL	**FVIPAGYPI**	**FVPYYNVNA**	**YVLNGSAWF**	**YVAFKTNDI**	**YKFLVPPPQ**
B part 4						
Lup a 1		433–442				
Lup a 4						
Lup a alpha conglutin	411–418		432–444	445–452	493–501	542–550
Lup an 3.0101	88.88% (FPILRWLGL) 413–421		434–442	447–455	495–503	544–552
Ara h 3			77% (FVPHYNTNA) 404–412			
Ara h 3.0201			77% (FVPHYNTNA) 454–465			

Allergen name	T-cell epitope
	FLLAAHAS
C.	
Range of amino acids occupied by T-cell epitopes joint over peanut	
Ara d 2	13–20
Ara h 2	13–21
Ara h 2.0101	13–21
Ara h 2.0201	13–21
Ara h 2.0202	13–21

This table lists the T -cell epitopes shared on at least two occasions by different species, describing the range of amino acids in which they are located and the percentage of identity with the epitope in the case in which identity is not exact.

Table 7.
Range of amino acids occupied by T-cell epitopes joint over legumes.

Gly m 5, *Gly m 5.0301*, and *Lup a 1* have shown that the IgE epitopes found on these proteins are part of the functional barrel domains of these proteins. In *Lup a 1* protein, two epitopes are located in the Cupin_1.1 domain, which is not affected by post-translational modifications; soybean proteins *Gly m 5* contain an IgE-epitope inside the *Cupin_1* domain, moreover *Gly m 5.0301* also contains the same epitope in the same region and in different positions having no modifications. However, *Gly m 5.0301* does contain epitopes directly affected by glycosidation, within the structural *Cupin_1* domain, an epitope at position 351 (**Table 5A, 6A** and **9A**).

The clinically proven epitopes found in the sequence analysis allowed us to observe how many and to what extent IgE epitopes are shared between proteins of different species and to assess potential cross-reactivity. According to the results, some of the

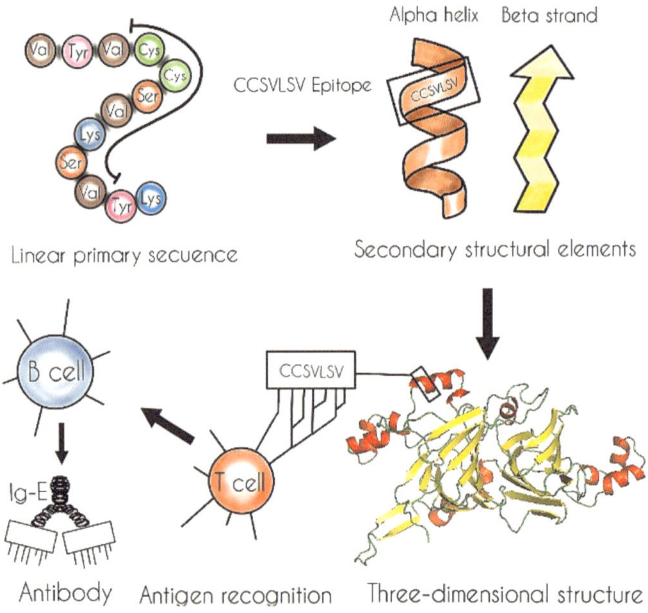

Figure 6.
Summary of the epitope recognition process.

Lupinus angustifolius			***Lupinus albus***		
Allergen name	**Based on amino acid composition**	**Based on dipeptide composition**	**Allergen name**	**Based on amin oacid composition**	**Based on dipeptide composition**
A.					
Prediction of Lupinus allergenic character					
Lup an 1	Potential allergen	Potential allergen	Lup a 1	—	—
Lup an 1.0101	—	—	Lup a 4	Potential allergen	Potential allergen
Lup an 3	Potential allergen	Potential allergen	Lup a alpha conglutin	—	—
Lup an 3.0101	Potential allergen	Potential allergen	Lup a delta conglutin	Potential allergen	Potential allergen
Lup an alpha conglutin	—	—	Lup a gama conglutin	—	—
Lup an delta conglutin	Potential allergen	Potential allergen	***Lupinus luteus***		
Lup an gamma conglutin	—	—	Lup l 4	Allergen	Potential allergen

Pisum sativum			***Glycine max***		
Allergen name	**Based on amino acid composition**	**Based on dipeptide composition**	**Allergen name**	**Based on amino acid composition**	**Based on dipeptide composition**
B.					
Prediction of pea and soy allergenic character					
Pis s 2	Potential allergen	Allergen	Gly m 5	Allergen	Allergen
Pis s 3	Potential allergen	Potential allergen	Gly m 5.0301	Allergen	Allergen
Pis s 3.0101	Potential allergen	Potential allergen	Gly m8	Allergen	Allergen
Pis s 6	Potential allergen	Potential allergen	Gly m 8.0101	Allergen	No allergen
Pis s aglutin	Potential allergen	Potential allergen			
Pis s albumin	Potential allergen	Potential allergen			

Cicer arietinum			***Arachis hypogaea***		
Allergen name	**Based on amino acid composition**	**Based on dipeptide composition**	**Allergen name**	**Based on amino acid composition**	**Based on dipeptide composition**
C.					
Prediction of chickpea and peanut allergenic character					
Cic a 1	—	—	Ara h 1	Allergen	Allergen
Cic a 3	Potential allergen	Allergen	Ara h 1.0101	Allergen	Allergen

Cicer arietinum			*Arachis hypogaea*		
Allergen name	**Based on amino acid composition**	**Based on dipeptide composition**	**Allergen name**	**Based on amino acid composition**	**Based on dipeptide composition**
Cic a 4	Potential allergen	Potential allergen	Ara h 2	—	—
Cic a 6	—	—	Ara h 2.0101	—	—
Arachis duranensis			Ara h 2.0201	—	—
Ara d 2	—	—	Ara h 2.0202	—	—
Ara d 6	—	—	Ara h 3	—	—
Arachis ipaensis			Ara h 3.0201	—	—
Ara i 2.0101	—	—	Ara h 5	Potential allergen	Potential allergen
Ara i 6.0101	—	—	Ara h 5.0101	Potential allergen	Potential allergen
Arachis hypogaea			***A. hypogaea***		
Allergen name	**Based on amino acid composition**	**Based on dipeptide composition**	**Allergen name**	**Based on amino acid composition**	**Based on dipeptide composition**
D.					
Prediction of peanut allergenic character					
Ara h 6	—	—	Ara h 11.0101	—	—
Ara h 6.0101	—	—	Ara h 11.0102	—	—
Ara h 7.0101	Allergen	—	Ara h 13.0102	Allergen	Allergen
Ara h 7.0201	—	—	Ara h 14.0101	—	—
Ara h 7.0301	—	—	Ara h 14.0102	—	—
Ara h 8	Potential allergen	Potential allergen	Ara h 14.0103	—	—
Ara h 8.0101	Potential allergen	Potential allergen	Ara h 15.0101	Allergen	Allergen
Ara h 8.0102	Potential allergen	Potential allergen	Ara h 16	Allergen	—
Ara h 9.0101	Allergen	Allergen	Ara h 17	Potential allergen	Allergen
Ara h 10.0101	—	—	Ara h aglutin	Potential allergen	—

The table summarizes the predictions about the allergenic potential of proteins based on the amino acid and peptide composition. The signal (−) means that the protein has clinically proven epitopes.

Table 8.
Allergenic legume character prediction.

candidate species and proteins for cross-reactivity with soybean (*G. max*) are the peanut (*A. hypogaea*) with three IgE epitopes commonly shared; lupin (*L. albus*) with one epitope in common (**Table 9A**). These findings are supported by bibliographic

Allergen name	**IgE epitopes**					
	HRIFLADKD	**NNFGKLFEVK**	**SYLQEFSRNT**	**ELHLLGFGIN**	**KDLAFPGSGE**	**RRYTARLKEG**
A.						
IgE epitopes shared between different legume species: Glycine max (Gly m), Lupinus albus /Lup a), Arachis hypogaea (Ara h), and Cicer arietinum (Cic a)						
Gly m 5	70% 415-QRNFLAGEKD	70% 297-NNFGKFFEIT	70% 217-SYLQGFSHNI			
Gly m 5.0301	70% 418-QRNFLAGEKD	70% 300-NNFGKFFEIT	70% 220-SYLQGFSHNI			
Lup a 1			70% 286-SYFSGFSRNT	80% 483-NLRLLGFGIN	70% 517-KELTFPGSAE	80% 456-RRYSARLSEG
Lup an 1.0101				80% NLRLLGFGIN	70% KELTFPGSIE	
Ara h 1	100% HRIFLADKD	90% NNFGRLFEVK	90% SYQGFSRNT	100% ELHLLGFGIN	100% KDLAFPGSGE	100% RRYTARLKEG
Ara h 1.0101	100% HRIFLADKD	100% NNFGKLFEVK	100% SYLQEFSRNT	100% ELHLLGFGIN	100% KDLAFPGSGE	100% RRYTARLKEG
Cic a 1				80% DLFLLGFGIN	70% KEVAFPGSAE	

Allergen name	**IgE epitopes**					
	GNIFSGFTPEFLEQA	**IETWNPNNQEFECAG**	**DRRCQSQLER**	**HASARQQWEL**	**KIQRDEDS**	**KRELRNL**
B.						
IgE epitopes shared between different legume species: Lupinus albus (Lup a), Lupinus angustifolius (Lup an), Arachis duranensis (Ara d), Arachis hypogaea (Ara h), and Cicer arietinum (Cic a)						
Lup a alpha conglutin	66.67% GNVLSGFDDEFLEEA	73.34% IETWNPKNDELRCAG				
Lup an alpha conglutin	66.67% GNVLSGFNDEFLEEA	73.34% IETWNPKNDQLRCAG				
Ara d 2			100% DRRCQSQLER	100% HASARQQWEL	100% KIQRDEDS	100% KRELRNL
Ara d 6						85.71% KRELMNL
Ara h 2			100% DRRCQSQLER	100% HASARQQWEL	100% KIQRDEDS	100% KRELRNL
Ara h 2.0101			70% DRRCQSQLER	100% HASARQQWEL	100% KIQRDEDS	100% KRELRNL
Ara h 2.0201			100% DRRCQSQLER	100% HASARQQWEL	100% KIQRDEDS	100% KRELRNL
Ara h 2.0202			100% DRRCQSQLER	100% HASARQQWEL	100% KIQRDEDS	100% KRELRNL
Ara h 3	86.67% GNIFSGFTSEFLAQA	100% IETWNPNNQEFECAG				
Ara h 3.0201	100% GNIFSGFTPEFLEQA	100% IETWNPNNQEFECAG				
Ara h 7.0201						85.71% ERELRNL
Cic a 6	73.33% GNIFSGFKRDFLEDA	73.33% IETWNPSNKQFACAG				

Allergen name	IgE epitopes						
	LQGRQQ	LRPCEQHLMQ	QRCDLDVE	QWELQGDR	RDPYSP	RDPYSP	SQDPYSPS
C.							
IgE epitopes shared between different legume species: Arachis duranensis (Ara d) and Arachis hypogaea (Ara h)							
Ara d 2	100% LQGRQQ	100% LRPCEQHLMQ	100% QRCDLDVE	100% QWELQGDR	100% RDPYSP	83.33% RDPYSP	100% SQDPYSPS
Ara d 6		80% LKPCEQHIMQ	87.5% QRCDLDVS				
Ara h 2	100% LQGRQQ	100% LRPCEQHLMQ	87.5% QRCDLEVE	100% QWELQGDR	100% RDPYSP	83.3% QDPYSP	100% SQDPYSPS
Ara h 2.0101	100% LQGRQQ	100% LRPCEQHLMQ	87.5% QRCDLEVE	100% QWELQGDR	100% RDPYSP	83.3% QDPYSP	100% SQDPYSPS
Ara h 2.0201	100% LQGRQQ	100% LRPCEQHLMQ	87.5% QRCDLEVE	100% QWELQGDR	100% RDPYSP	83.3% QDPYSP	100% SQDPYSPS
Ara h 2.0202	100% LQGRQQ	100% LRPCEQHLMQ	87.5% QRCDLEVE	100% QWELQGDR	100% RDPYSP	83.3% QDPYSP	100% SQDPYSPS
Ara h 6.0101			87.5% QRCDLDVS				
Ara h 7.0201			70% LRPCEEHIRQ				
Ara h 7.0301		70% LRPCEEHIRQ					
D.							
IgE epitopes shared between different legume species: Arachis duranensis (Ara d) and A. hypogaea (Ara h)							
Ara d 2	100% LQGRQQ	100% LRPCEQHLMQ	100% QRCDLDVE	100% QWELQGDR	100% RDPYSP	83.33% RDPYSP	100% SQDPYSPS
Ara d 6		80% LKPCEQHIMQ	87.5% QRCDLDVS				
Ara h 2	100% LQGRQQ	100% LRPCEQHLMQ	87.5% QRCDLEVE	100% QWELQGDR	100% RDPYSP	83.3% QDPYSP	100% SQDPYSPS
Ara h 2.0101	100% LQGRQQ	100% LRPCEQHLMQ	87.5% QRCDLEVE	100% QWELQGDR	100% RDPYSP	83.3% QDPYSP	100% SQDPYSPS
Ara h 2.0201	100% LQGRQQ	100% LRPCEQHLMQ	87.5% QRCDLEVE	100% QWELQGDR	100% RDPYSP	83.3% QDPYSP	100% SQDPYSPS
Ara h 2.0202	100% LQGRQQ	100% LRPCEQHLMQ	87.5% QRCDLEVE	100% QWELQGDR	100% RDPYSP	83.3% QDPYSP	100% SQDPYSPS
Ara h 6.0101			87.5% QRCDLDVS				
Ara h 7.0201		70% LRPCEEHIRQ					
Ara h 7.0301		70% LRPCEEHIRQ					

This table summarizes the IgE epitopes clinically confirmed in different species, and the accuracy percentage of these epitopes found according to the protein sequence.

Table 9.
IgE epitopes shared between different legume species.

reports [38]. It is also found that *L. albus* shares four epitopes with *A. hypogea* and *L. angustifolius* (**Table 9A**), and other two with *A. hypogea*. Looking at other cases, it is observed that in close species such as peanut, species such as *A. duranensis* and *A.*

Allergen name	IgE epitopes			
	DITNPINLRE	**KESHFVSARP**	**EQEERGQRRW**	**VTVRGGLRILSPDRK**
IgE epitopes shared only by same legume species: Arachis hypogaea (Ara h)				
Ara h 1	90% DITNPINLRD	90% RESHFVSARP	90% EQEERGQRR	
Ara h 1.0101	100% DITNPINLRE	100% KESHFVSARP	100% EQEERGQRRW	
Ara h 3				93,345% VTCRGGLRILSPDRK
Ara h 3.0201				86.67% VTVRGGLRILSPDRK

This table summarizes the IgE epitopes clinically confirmed in different species, and the accuracy percentage of these epitopes found according to the protein sequence.

Table 10.
IgE epitopes shared only by same legume species.

hypogea shared ten common epitopes (**Table 9B, C**), similarly to *Lupinus* finding four epitopes in common (**Table 9A, B**).

In addition, shared T-cell epitopes have been found among species that do not include soybean such as *L. albus* and *L. angustifolia* (**Table 9**: AB), but not found in *L. luteus*; *A. hypogaea* (**Table 9A-D**), and *A. duranensis* (**Table 9B-D**); *C. arietinum* (**Table 9A, B**); and *P. sativum* (**Table 9A**). These epitopes have been identified as relevant epitopes in previous studies on sensitizations between allergens of different species with similar structure and sequence leading to the development of allergic cross-reactions [38, 39].

An interesting fact is that different isoforms of the same protein may or may not present the same IgE epitope and, in the case of having it, it does not necessarily have the same degree of similarity. Establishing a relationship with the information obtained in the alignments, we can conclude that the small differences observed in the sequence between isoforms of the same protein can be key to conformation and epitopes presence (**Table 10**).

4. Conclusions

This chapter presented a study of functional and allergenic features of legume seed proteins.

Analysis of allergenic legume proteins legume as well as all available isoforms allowed for extracting shared epitopes that can be linked to cross-reactivity processes among the eight studied species (*G. max, A. hypogaea, L. albus, L. angustifolius, A. duranensis, C. arietinum, P. sativum,* and *L. culinaris)*. Shared epitopes were not found with soybean or with the rest of the legume allergens examined from *A. duranensis*.

Small differences in the amino acid sequences (less than 1%) of the same allergen isoforms implied important changes in epitopic conformation and sequences of T-cell and IgE recognizable epitopes. Small differences in amino sequences of isoforms from the same inferred changes over 2D and 3D structure conformation that may affect functional protein domains. Post-translational modifications allowed identification of possible phosphorylation, glycosylation, carboxylation, methylation, nitrosylation, and nitration sites in protein functional domains, near or directly located in different type of epitopes with potential influence in allergenic response.

Primary sequence alignments together with three-dimensional protein modeling allowed to study the conservation of proteins as conglutin gamma proteins among different *Lupinus*. species, assessing also their potential allergenicity.

The changes described close to the sequence or related to spatial distribution of the epitopes may involve potential alterations on protein allergenicity.

Obtaining reliable clinical data on legume allergies in developing countries could be helpful in clarifying whether the increase in food allergies is actually due to poor dietary habits and increasing industrialization processes.

Further studies on the characterization of more allergenic proteins, including isoforms of major allergens already described, not only sequential but also three-dimensional conformational epitopes, can be a great advancement for the prevention of cross-reactivity and the improvement of knowledge of allergies produced by legumes, which in turn could promote the introduction of this food as a substitute for other foods of lower nutritional quality and with greater environmental impact.

Acknowledgements

This study has been partially funded by The Spanish Ministry of Economy, Industry and Competitiveness through the grants Ref.: RYC-2014-2016,536 (Ramon y Cajal Research Program) to JCJ-L; and Ministry of Health and Families, Andalusian government. Funding for I + D + i in biomedical research and health sciences in Andalusia, grant Ref.: PI-0450-2019.

Conflicts of interest

The authors have declared that no competing interests exist.

Abbreviations

LTP	Lipid Transfer Protein
3D	three-dimensional
PR	proteins Pathogenesis-related proteins

Author details

Andrea Roman-Mateo[1], Esther Rodriguez-de Haro[1], Jose M. Berral-Hens[1], Sonia Morales-Santana[2] and Jose C. Jimenez-Lopez[1,3*]

1 Department of Biochemistry, Cell and Molecular Biology of Plants, Estación Experimental del Zaidín, Spanish National Research Council (CSIC), Granada, Spain

2 Proteomic Research Department, San Cecilio University Hospital, Biosanitary Research Institute of Granada (Ibs.GRANADA), Granada, Spain

3 The UWA Institute of Agriculture and School of Agriculture and Environment, The University of Western Australia Perth, WA Australia

*Address all correspondence to: josecarlos.jimenez@eez.csic.es

References

[1] Myles IA. Fast food fever: Reviewing the impacts of the Western diet on immunity. Nutrition Journal. 2014;**13**. Article number 61:2-4

[2] Tang MLK, Mullins RJ. Food allergy: Is prevalence increasing? Internal Medicine Journal. 2017;**47**:256-261

[3] Gray CL, Levin ME, du Toit G. Ethnic differences in peanut allergy patterns in south African children with atopic dermatitis. Pediatric Allergy and Immunology. 2015;**26**: 721-730

[4] Rani K, Sharma P, Kumar S,Wati L, Kumar R, Gurjar DS, et al. Legumes for Sustainable Soil and Crop Management. In: Meena R, Kumar S, Bohra J, Jat M. (eds) Sustainable Management of Soil and Environment. Singapore: Springer; DOI: 10.1007/978-981-13-8832-3_6

[5] Stagnari F, Maggio A, Galieni A, et al. Multiple benefits of legumes for agriculture sustainability: An overview. Chemical and Biological Technologies in Agriculture. 2017;**4**:2

[6] Hildebrand HV, Arias A, Simons E, Povolo B, Janet Rothney MLIS, Protudjer JLP. Adult and pediatric food allergy to chickpea, pea, lentil, and lupine: A scoping review. The Journal of Allergy and Clinical Inmmunology. 2020;**9**(1):290-301

[7] Scott H, Sicherer MD. Clinical implications of cross-reactive food allergens. Journal of Allergy and Clinical Immunology. 2001;**108**(6):881-890

[8] Jappe U, Vieths S. Lupine, a source of new as well as hidden food allergens. Molecular Nutrition Food Research. 2010;**54**:113-126

[9] San Ireneo MM, Ibáñez MD, Sánchez J-J, Carnés J, Fernández-Caldas E. Clinical features of legume allergy in children from a Mediterranean area. Annals of Allergy, Asthma and Inmunology. 2008;**101**(2):179-184

[10] Alvarado MI, Pérez M. Study of food allergy on Spanish population. Allergologia et Immunopathologia. 2006;**34**(5):185-193

[11] Richard C, Jacquenet S, Sergeant P, Moneret-Vautrin DA. Cross-reactivity of a new food ingredient, dun pea, with legumes, and risk of anaphylaxis in legume allergic children. European Annals of Allergy and Clinical Immunology. 2015;**474**(4):118-125

[12] Nwaru BI, Hickstein L, Panesar SS, Muraro A, Werfel T, Cardona V. On behalf of the EAACI Food Allergy and Anaphylaxis Guidelines Group. The epidemiology of food allergy in Europe: A systematic review and meta-analysis. Allergy. 2014;**69**:62-75

[13] Tham EH, Leung DYM. How different parts of the world provide new insights into food allergy. Allergy Asthma Immunol research. 2018;**10**(4):290-299

[14] Gaiaschi R, Beretta P, Fiocchi C, Velonà P, Galli U. Cross-reactivity between milk proteins from different animal species. Clinical & Experimental Allergy. 2001;**9**(7):997-1004

[15] Lallès JP, Peltre G. Lead review article biochemical features of grain legume allergens in humans and animals. Nutrition Reviews. 1996;**54**(4):101-107

[16] Mills ENC, Jenkins J, Marigheto N, Belton PS, Gunning AP, Morris VJ. Allergens of the Cupin superfamily.

Biochem Society Transactions. 2002; **30**(6):925-929

[17] Asero R, Mistrello G, Roncarolo D, de Vries S, Gautier M, Ciurana CL, et al. Lipid transfer protein: A pan-allergen in plant-derived foods that is highly resistant to pepsin digestion. International Archive of Allergy Immunology. 2001;**124**:67-69

[18] Wang D, Liu D, Yuchi J, He F, Jiang Y, Cai S, et al. MusiteDeep: A deep-learning based webserver for protein post-translational modification site prediction and visualization. Nucleic Acids Research. 2020;**48**(1):140-146

[19] Kattan JD, Sampson HA. Clinical reactivity to soy is best identified by component testing to Gly m 8. Journal Allergy and Clinical Immunoly. 2015; **3**(6):970-972

[20] Ueberham E, Spiegel H, Havenith H, Rautenberger P, Lidzba N, Schillberg S, et al. Simplified tracking of a soy allergen in processed food using a monoclonal antibody-based Sandwich ELISA targeting the soybean 2S albumin Gly m 8. Journal of Agricultural and Food Chemistry. 2019;**67**(31):8660-8667

[21] Maruyama N, Sato S, Cabanos C, Tanaka A, Ito K, Ebisawa M. Gly m 5/Gly m 8 fusion component as a potential novel candidate molecule for diagnosing soya bean allergy in Japanese children. Clinical and Experimental Allergy. 2018; **48**:1726-1734

[22] CODEX Alimentarius Commission. Proposed draft annex on the assessment of possible allergenicity of the draft guideline for the conduct of food safety assessment of foods derived from recombinant DNA plants. Joint FAO/ WHO Food Standard Program. Appendix IV. 2003:57-60

[23] Midun E, Radulovic S, Brough H, Caubet J-C. Recent advances in the management of nut allergy. World Allergy Organization Journal. 2021; **14**(1):1939-4551

[24] Bublin M, Breiteneder H. Cross-reactivity of Peanut allergens. Current Allergy and Asthma Reports. 2014;**14**: 426

[25] McGinty JW, Marré ML, Bajzik V, et al. T cell epitopes and post-translationally modified epitopes in type 1 diabetes. Currently Diabetes Repport. 2015;**15**:90

[26] Narayanan A, Jacobson MP. Computational studies of protein regulation by post-translational phosphorylation. Current Opinion in Structural Biology. 2009;**19**(2):156-163

[27] Deribe Y, Pawson T, Dikic I. Post-translational modifications in signal integration. Nature Structural & Molecular Biology. 2010;**17**:666-672

[28] Mueller GA, Maleki SJ, Johnson K, Hurlburt BK, Cheng H, Ruan S, et al. Identification of Maillard reaction products on peanut allergens that influence binding to the receptor for advanced glycation end products. Allergy. 2013;**68**:1546-1554

[29] Kurotani A, Tokmakov AA, Kuroda Y, Fukami Y, Shinozaki K, Sakurai T. Correlations between predicted protein disorder and post-translational modifications in plants. Bioinformatics. 2014;**30**(8): 1095-1103

[30] Astier J, Rasul S, Koen E, Manzoor H, Besson-Bard A, Lamotte O, et al. S-nitrosylation: An emerging post-translational protein modification in plants. Plant Science. 2011;**181**(5): 527-533

[31] Corpas FJ, Begara-Morales JC, Sánchez-Calvo B, Chaki M, Barroso JB. Nitration and S-Nitrosylation: Two Post-translational Modifications (PTMs) Mediated by Reactive Nitrogen Species (RNS) and Their Role in Signalling Processes of Plant Cells. In: Gupta K, Igamberdiev A. (eds) Reactive Oxygen and Nitrogen Species Signaling and Communication in Plants. Signaling and Communication in Plants, vol 23. Cham: Springer. 2015. DOI: 10.1007/978-3-319-10079-1_13

[32] Dunwell JM, Culham A, Carter CE, Sosa-Aguirre CR, Goodenough PW. Evolution of functional diversity in the Cupin superfamily. Trends in Biochemical Sciences. 2001;**26**(12): 740-746

[33] Jimenez-Lopez JC, Rodríguez-García MI, Alché JD. Analysis of the Effects of Polymorphism on Pollen Profilin Structural Functionality and the Generation of Conformational, T- and B-Cell Epitopes. PLOS ONE. 2013;**8**(10): e76066

[34] Svobodova M, Mairal T, Nadal P, Bermudo MC, O'Sullivan CK. Ultrasensitive aptamer based detection of β-conglutin food allergen. Food Chemistry. 2014;**165**:419-423

[35] Nadal P, Svobodova M, Mairal T, O'Sullivan CK. Probing high-affinity 11-mer DNA aptamer against Lup an 1 (β-conglutin). Analytical and Bioanalytical Chemistry. 2013;**405**(29):9343-9349

[36] Cabanos C, Katayama H, Tanaka A, Utsumi S, Maruyama N. Expression and purification of peanut oleosins in insect cells. Protein Journal. 2011;**30**(7): 457-463

[37] Ramos ML, Fleming G, Chu Y, Akiyama Y, Gallo M, Ozias-Akins P. Chromosomal and phylogenetic context for conglutin genes in Arachis based on genomic sequence. Molecular Genetics and Genomics. 2006;**275**(6):578-592

[38] Cabanillas B, Jappe U, Novak N. Allergy to Peanut, Soybean, and Other Legumes: Recent Advances in Allergen Characterization, Stability to Processing and IgE Cross-Reactivity. Molecular Nutrition and Food Research. 2018;**62**(1)

[39] Verma AK, Kumar S, Das M, et al. A comprehensive review of legume allergy. Clinical Reviews in Allergy & Immunology. 2013;**4**:30-46